ACOUSTIC SURFACE WAVE AND ACOUSTO-OPTIC DEVICES

Edited by Thomas Kallard

optosonic press

ACOUSTIC SURFACE WAVE AND ACOUSTO-OPTIC DEVICES

(STATE OF THE ART REVIEW, Vol. 4)

Published by Optosonic Press
Box 883, Ansonia Post Office
New York, N.Y. 10023

Additional copies may be procured
by addressing the publisher.

Library of Congress Catalog Card No. 77-160226

International Standard Book Number: 0-87739-003-7

PRINTED IN THE UNITED STATES OF AMERICA

EDITOR'S PREFACE

Optosonic Press takes pleasure in offering this fourth volume in the STATE OF THE ART REVIEW Series.

The purpose of this volume is to bring to the attention of the readership the work being done in the field of acoustic surface waves and acousto-optic interactions. Many of the recently discovered effects and device developments are described. The 16 papers contained in this volume are of particular interest to microwave engineers and electro-optical systems designers.

The review articles, state-of-the-art research reports, the bibliography and the patent listing, in combination, serve to put into the hands of the reader an up-to-date compilation of highly specialized information. It is hoped, with such a ready source of reference, the usually high percentage of unintentionally duplicated R & D work can be reduced to a minimum.

The Editor wishes to thank the authors for their time and effort in producing the excellent papers. Our gratitude is extended to the cooperating companies and institutions who allowed the authors time to develop their papers.

New York, N.Y.
September, 1971

Thomas Kallard
for OPTOSONIC PRESS

CONTENTS

ACOUSTIC SURFACE WAVES

SURFACE ELASTIC WAVES --- A REVIEW

Dr. Edgar A. Kraut
Science Center
North American Rockwell Corporation
Thousand Oaks, California 91360

Introduction

Surface elastic waves have wavelengths long compared to interatomic dimensions and characteristically exhibit a decay in amplitude with distance away from the surface along which they travel. This very general kind of wave motion occurs in a wide variety of physical systems. For example, seismic surface waves traveling in the earth's crust have periods of up to 100 seconds, and wavelengths on the order of 400 km while the same kind of elastic surface wave can be electrically excited on the polished surface of a piezoelectric crystal at sub-nanosecond periods with wavelengths less than a micron.[2]

There are several different kinds of surface elastic waves. These differ mainly according to whether the surface to which they are confined is a stress free boundary of an elastic solid or is an interface separating two physically dissimilar media. All elastic solids can support surface waves, independently of whether the solid is isotropic or anisotropic, homogeneous or inhomogeneous. However, anisotropic solids exhibit a richer variety of surface wave phenomena, such as weak or oscillatory decay with depth along certain symmetry directions,[3-9] decoupling of the elctric from the elastic field for surface wave propagation in certain special directions in piezoelectric crystals;[10-12] and the existence of a class of surface waves in piezoelectric crystals with no nonpiezoelectric counterpart at all.[13,14]

Classification of Surface Elastic Waves

Frequency-wavenumber dispersion, variation of phase and group velocity with wavelength, rate and type of amplitude decay with depth, and particle displacement polarization are commonly used to distinguish a classify types of surface elastic waves. Extended discussions of the details may be found in recent books by Victorov[15], Tiersten[16], Brekhovskikh[17], Ewing et al.[18], White[19], Holland and Eernisse[20], and Musgrave[21], and in recent review articles by Farnell[5], White[22], Dransfeld and Salzmann[23], de Klerk[24], Sabine and Cole[25], Smith[26] and Kraut[27]. A general bibliography of microwave ultrasonics has been compiled by Smith and Damon[28].

The polarization of surface elastic waves is conveniently described with respect to the sagittal plane[29]. This plane is defined as the plane determined by the normal to the surface and the propagation or $\underline{k}$ vector of the propagating surface vave. The simplest and most important kind of surface elastic wave was discovered by Lord Rayleigh[30] in 1885. Rayleigh waves, or acoustic surface waves as the electrical engineers often call them, propagate along the stress free boundary surface of a semi-infinite elastic half space. In the absence of mechanical dissipation, Rayleigh waves are nondispersive. They travel at a speed independent of the wavelength. This speed is characteristic of a material and is determined by its elastic moduli and density. The range of Rayleigh wave speeds encountered in most solids lies between 10^5 and 6×10^5 cm/sec, i.e., about five orders of magnitude lower than the speed of light. As a consequence, a Rayleigh wave at a given frequency has a wavelength which is about five orders of magnitude shorter than the corresponding wavelength for an electromagnetic wave of the same frequency. Similarly, to delay an electromagnetic signal by one microsecond requires about $3 \cdot 10^5$ mm of coaxial cable but the same delay for Rayleigh waves on quartz can be provided by only about 4 mm of quartz. Considerations such as these are responsible for the current flurry of interest in Rayleigh waves prevalent in the device physics community.

In isotropic solids, Rayleigh waves have no component of particle motion traverse to the sagittal plane. Within the sagittal plane, there is a component of particle motion normal to the free surface and a component parallel to the $\underline{k}$ vector. As the Rayleigh wave propagates, individual particles at and below the free surface execute elliptical orbits in the sagittal plane with the major axis of the ellipse perpendicular to the free surface. At and near the free surface, the particle motion is retrograde elliptical. The shape of the ellipse changes with depth because the major and minor axis change with depth. At a depth of about 0.18 wavelength the particle motion is linearly polarized and at greater depths it becomes elliptical prograde.

Stoneley[31] extended Rayleigh's work[30] to cubic crystals and shortly thereafter Synge[3,4] considered Rayleigh waves in arbitrary anisotropic nonpiezoelectric media. He concluded that for a given circular frequency ω such waves either do not exist or travel only in certain directions in the free surface, with certain corresponding speeds. Gazis et al.[32] calculated surface wave velocities for the free surface in the (001) plane of many cubic crystals and claimed that for some, such as aluminum and copper, surface waves "do not exist in a range of directions about the (110) direction although a surface wave is always possible in the (110) direction itself." Buchwald and Davis[33] state categorically that "Surface waves in anisotropic media are possible only if the free plane is a plane of symmetry of the material. In media with cubic symmetry the only possible cases of surface wave propagation are on the (001) or (110) planes." Moreover, their calculations in the (001) plane of aluminum, iron, lead, and near the (110) axis in the (110) plane of rocksalt. Similar computations for LiF and Cu published by Tursonov[34] show directions near (110) as being forbidden. Lim and Farnell[6,7] carried out a detailed computer search for forbidden directions of elastic surface wave propagation in anisotropic crystals. They state[6] that "an extensive search of propagation directions on many planes in many crystals has not revealed any forbidden directions in that there always appears to be a surface wave solution which satisfies the free surface boundary condition and which is unattenuated in the direction of propagation. Near certain specific directions the penetration of this wave below the surface becomes very deep. For certain ranges of direction there are second or pseudo-surface-wave solutions which in general involve a radiating component into the bulk. The latter component can be small enough to make the pseudo-surface-wave easily detectable.[8,9]" In view of the computational results of Lim and Farnell[5-7], Burridge has recently re-examined the theoretical question as to the directions in which Rayleigh waves may be propagated on crystals.[35] He proves that the assertion[33] that surface waves in anisotropic media are possible only if the free plane is a plane of material symmetry is incorrect. Moreover, he proves that if Rayleigh waves exist at all on any arbitrary plane in a totally anisotropic medium then there must be whole angular ranges of directions in which they propagate and not just the discrete directions suggested by Synge.[3]

In summary, the notions of discrete allowed directions, and forbidden ranges of directions for surface wave propagation in anisotropic media must be considered as incorrect. The forbidden ranges appear now to be those in which the penetration of the surface wave becomes relatively deep.[6] If the sagittal plane is not a symmetry plane, the particle motion may include three instead of just two orthogonal components. Amplitude decay with depth may be in the form of a damped oscillation with depth instead of purely exponential decay with depth. For certain isolated angles on some anisotropic substrates pseudo or leaky surface waves propagate with a velocity higher that that of the slowest transverse bulk wave. Such surface waves radiate bulk shear waves in to the interior of the solid and this causes the dissipation of the surface wave. In many directions the surface wave propagation vector and the direction of energy propagation are not colinear.[6] This non-parallelism of surface wave phase and group velocity is of considerable importance in the choice of crystal orientations for surface wave devices, as energy launched at one point may entirely miss a distant receiver because of this beam steering effect. In pure mode directions the surface wave phase and group velocities are colinear.

Surface Elastic Waves in Piezoelectric Solids

In piezoelectric solids electromagnetic as well as stress boundary conditions must be satisfied at a free surface. Since elastic wave speeds are about five orders of magnitude less than the speed of light, piezoelectric insulators are treated electrostatically.[22,23,29] The boundary conditions require continuity of the tangential electric field and the normal component of electric displacement at the boundary surface. Some important differences between surface elastic waves in piezolectric and non-piezoelectric solids are the following:[22]

1) The propagating surface wave in a piezoelectric solid is accompanied both inside and outside the solid by a traveling electric field.

2) Inclusion of piezoelectricity for a given solid stiffens the elastic constants and results in a higher wave velocity than for an elastically similar but non-piezoelectric solid.

3) Inclusion of piezoelectricity may cause an oscillatory decay of surface wave amplitude with depth.

4) The wave velocity now depends on piezoelectric, dielectric and elastic properties and the temperature dependence of the surface wave velocity is affected by the temperature dependences of those properties.

5) In piezoelectric solids electro-acoustic surface waves[13,14] can propagate, having displacements entirely transverse to the sagittal plane. Such surface waves have no counterpart in homogeneous isotropic media.

Excitation of Surface Elastic Waves

Lamb forst considered[36] the excitation of a disturbance in a semi-infinite isotropic elastic medium by the application of a vertical or horizontal impulse along a line in the free surface. Lamb's problem[36] is very famous among geophysicists and is considered by many to be the beginning of modern theoretical seismology. Lamb's problem was solved for anisotropic non-piezoelectric media possessing a six fold symmetry axis by Kraut[27,37] and independently by Abubakar.[38,39] Burridge has recently solved Lamb's problem[40] for an arbitrary surface in a completely anisotropic non-piezoelectric medium. It is clear from the foregoing investigations that although it is possible to discuss surface waves independently from bulk waves, the impulse response or "Green's function" of a semi-infinite elastic medium contains both bulk (continuous spectrum) and surface wave (discrete spectrum) contributions. Because bulk waves radiate into the volume of a solid while the surface wave remains confined to the free surface, the far field of an impulse source is dominated at the free surface by the surface elastic wave contribution.

The excitation of surface elastic waves in isotropic and anisotropic non-piezoelectric media is usually accomplished by the application of a localized vertical or horizontal impulse at the free surface[19], although other steady state excitation methods are also available[15,23]. For piezoelectric crystals, modern photolithographic techniques[41] favor the use of photographically produced thin metallic film grating patterns for surface wave excitation.[22-25] In fact, advances in microelectronic photolithography and thin film techniques combined with the recent availability of large single crystals of strong piezoelectrics like $LiNbO_3$ are responsible, to a large extent, for renewed interest in surface elastic waves.

A typical grating consists of a pattern of metallic fingers. If these are numbered consecutively starting from one end, then all the even fingers are connected together and run to one side of an AC signal generator while all the odd fingers are similarly connected together and run to the other side of the generator. If the grating spacing is uniform, then the center to center distance between a pair of even or odd fingers determines the wavelength of the elastic surface wave which will be excited. The AC generator is tuned to the frequency corresponding to the elastic surface wave velocity (a material parameter) divided by the spacing between a pair of adjacent even or odd fingers. The electrical properties such as radiation resistance, transduction efficiency, and electrical equivalent circuit which characterize a metallic grating on a piezoelectric substrate have been the subject of recent intensive investigation[42-51] particularly in the electrical engineering community. The closely related problem of determining the best surface in a piezoelectric crystal and the best direction in that surface to use in order to maximize conversion of electrical input power into surface elastic wave energy has also been solved for a number of different crystals.[10,52-55]

Attenuation, Scattering and Temperature Dependence of Surface Elastic Waves

The variation of surface wave speed with temperature is an important parameter in delay line applications, and for quartz a surface having particularly favorable temperature and coupling characteristics is known.[56] Surface wave temperature coefficient data for lithium tantalate is available[57] and the temperature dependence of the elastic constants of bismuth germanium oxide has also been measured.[53]

Absorption mechanisms for Rayleigh waves in solids have been reviewed by Dransfeld and Salzmann.[23] These consist of:

(a) Scattering from surface imperfections (scratches, polishing defects);

(b) Scattering from crystal defects (dislocations and grain boundaries in polycrystalline materials);

(c) Interactions with thermal phonons (surface and bulk phonons);

(d) Interactions with electrons in metals;

(e) Interactions with drifting carriers in semiconductors;

(f) Attenuation due to interaction with gases or liquids adjacent to a solid surface.

The attenuation of surface waves due to scattering has been examined theoretically by Steg and Klemens[58] and by Hudson.[59] Experimental attenuation measurements are reported in the literature[22,23,60] for a number of commonly used piezoelectric crystals. Typical numerical values for surface wave attenuation are[60] 1.1-1.78 dB per microsecond at 316 MHz in Y cut X propagating quartz.

Nonlinear Properties of Surface Elastic Waves

There exist several kinds of nonlinear surface elastic wave interactions in solids. Some of these are unique to piezoelectric crystals and involve the coupling of the elastic and dielectric properties of a crystal lattice by anharmonicities in interatomic forces. Rischbieter[61,62] has observed second-harmonic generation in elastic surface waves in aluminum and steel where the nonlinearity is due to both lattice anharmonicity and dislocations. Lopen has reported[63] second harmonic generation in an elastic surface wave in alpha quartz. In this case the nonlinearity in question can be explained as due to third order anharmonic elastic constants. Similar effects have been observed by Lean and coworkers.[63,64]

A different kind of nonlinear surface wave interaction unique to piezoelectric materials has been observed in lithium niobate by Svaasand[65] and by Luukkala and Kino[66], and in lead zirconate titanate (PZT) by Kraut, Lim and Tittmann.[67] In this kind of nonlinear interaction, a pair of oppositely propagating elastic surface waves with angular frequency and wave number (ω_1, k_1) and (ω_2, k_2) respectively, generate an electric field at frequency $(\omega_1 + \omega_2)$ with wave number $(k_1 - k_2)$ and an electric field at frequency $(\omega_1 - \omega_2)$ with wave number $(k_1 + k_2)$.[65] This effect has been used as the basis of surface wave signal processing devices which can convolve, cross correlate or fourier transform input electrical pulses in real time.[66,67]

Surface Waves in Layered Media

The developmentof heteroepitaxial thin film technology based on chemical vapor deposition (CVD) and RF sputtering techniques has led to the development of surface elastic wave amplifiers[68,69] dispersive delay lines[70,71] and waveguides.[72,73] Currently, some 60 heteroepitaxial combination is available for study.[74] For a single layer on a semi-infinite substrate[75,76] the fundamental symmetric mode of vibration, at wavelengths short compared to the layer thickness, reduces to an elastic surface wave traveling at the Rayleigh wave speed of the layer. At wavelengths long compared to the layer thickness, the fundamental symmetric mode reduces to a Rayleigh wave charavteristic of the substrate. For intermediate wavelengths the fundamental symmetric mode is dispersive. In addition to the fundamental symmetric mode there is a second fundamental mode called a Sezawa mode [77] which exists in layered media. Tolstoy and Usdin[78] explained the difference between the two modes. The Rayleigh modes correspond to symmetrical modes of a free plate modiefied by the contact of the lower surface to the semi-infinite elastic solid. The Sezawa modes correspond to a similar modification of the antisymmetrical free plate modes. The free plate modes themselves are called Lamb waves.[15] A single layer on a semi-infinite substrate also exhibits a second type of dispersive surface wave called a Love wave. The particle motion in a Love wave is purely transverse to the sagittal plane.

For a pair of isotropic semi-infinite elastic solids joined along an interface, for a limited range of material constants[80], a surface wave confined to the interface exists. This type of surface wave, called a Stoneley wave[81], can also exist at the interface of a single layer on a semi-infinite substrate.[82] Stoneley waves in anisotropic nonpiezoelectric media have been investigated in detail by Lim and Musgrave[83] and by Johnson.[84] Preliminary considerations regarding symmetry relations for Stoneley waves in anisotropic piezoelectric media have been published by Farnell.[12]

The computation of phase and group velocity in multilayered isotropic media has been recently reviewed by Schwab[85] and a discussion of similar procedures for multilayered anisotropic non-piezoelectric media has been given by Crampin.[86] For multilayered completely anisotropic piezoelectric media the corresponding computational results are not yet available, although Solie reports[87] the development of a computer program for calculating surface wave dispersion curves for the film-on-substrate problem where both materials may be arbitrarily anisotropic and piezoelectric. A matrix formulation of the same problem has been given by Kraut.[88] Similar formulations have been discussed in isotropic problems by Gilbert and Backus[89], and by Abramovici[90,91], and in anisotropic piezoelectric problems by Ingebrigtsen.[94]

Excitation of Elastic Waves in Layered Media

A definitive treatment of the impulsive excitation of elastic waves in an isotropic layer on a semi-infinite isotropic substrate was given by Newlands[82] in 1952. Since that time, the same problem has been treated from a normal-mode and leaky mode point of view[91,95,96], from a ray theory approach[97] and numerically by finite differences.[98] The same wealth of analytical and numerical detail is not yet available for anisotropic layered solids, piezoelectric or otherwise. The normal and leaky mode studies of layered isotropic solids[90,91,95,96] first emphasized the classification of elastic wave modes according to the behavior of the real and imaginary parts of their frequency versus wavenumber dispersion relations as opposed to the older classification based on the variation of phase and group velocity with wavelength. A detailed theory of the operation of an interdigital transducer on the free surface of or embedded within a layered piezoelectric solid is not yet available.

Elastic Surface Wave Devices

Radar and communications systems demand rapid acquisition and processing of vast quantities of data. Complex coded waveforms and wideband signals for missile guidance and aircraft flight control must be handled with high resolution. Delay lines are essential which can store these signals, filter them in the time of frequency domain, amplify, compress, expand, decode or fourier analize their waveforms. All of these signal processing functions can now be performed with elastic surface wave devices.[22,99-103] The basic elastic surface wave delay line shown in Figure 1 has an insertion loss and phase shift which varies with frequency so that the delay line acts as a frequency domain filter. Tancrell has recently published[103] a detailed discussion of the design and application of surface wave filters.

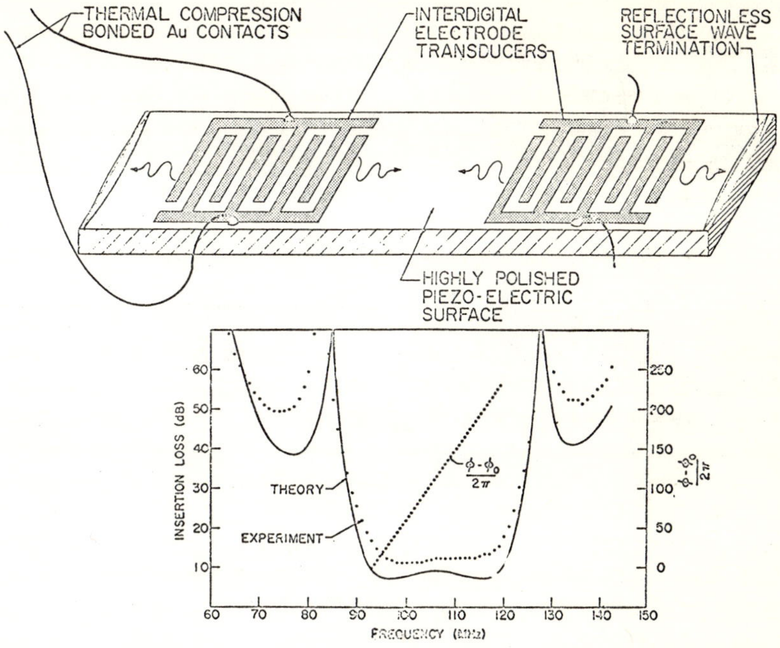

Fig. 1

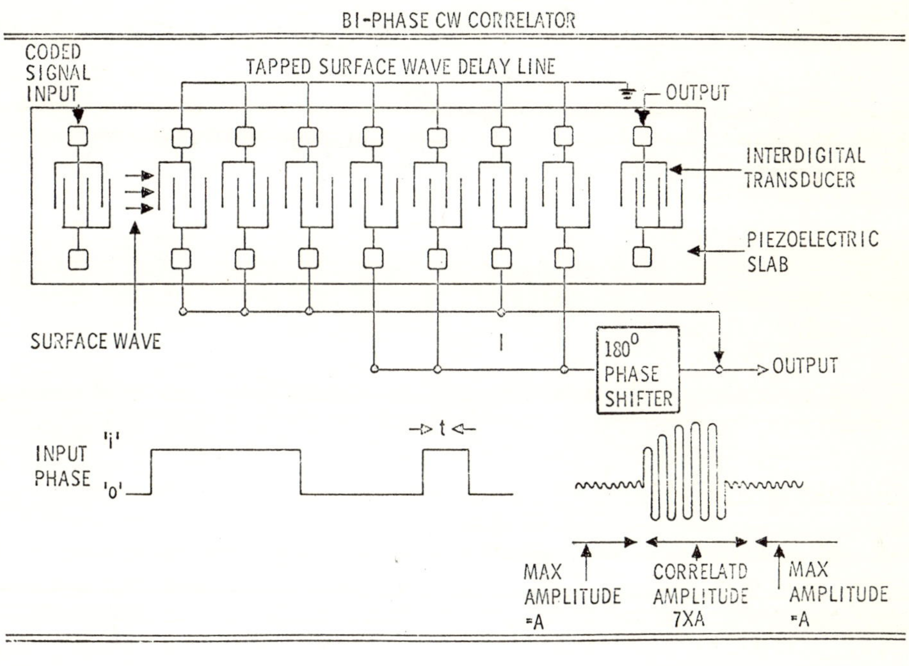

Fig. 2

Time domain filtering, signal correlation, decoding and encoding can be performed with a tapped surface wave delay line[101] such as that shown in Fig. 2. When a surface wave whose oscillations coincide with the spatial tap positions is in registration with the taps a correlation peak in the output from the summed taps is obtained.

ACOUSTIC SURFACE WAVE CONVOLVER

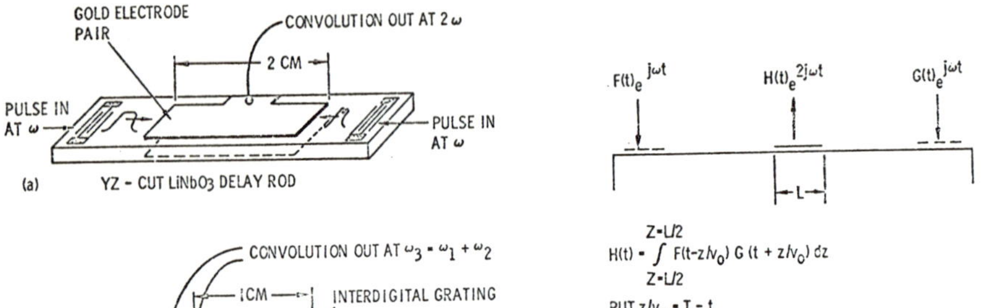

$$H(t) = \int_{Z=-L/2}^{Z=L/2} F(t - z/v_0)\, G(t + z/v_0)\, dz$$

PUT $z/v_0 = T - t$

THEN $H(t) = v_0 \int_{-\infty}^{\infty} G(T)\, F(2t - T)\, dT$

H(t) IS THE CONVOLUTION FUNCTION OF G (T) AND F(T)

Fig.3

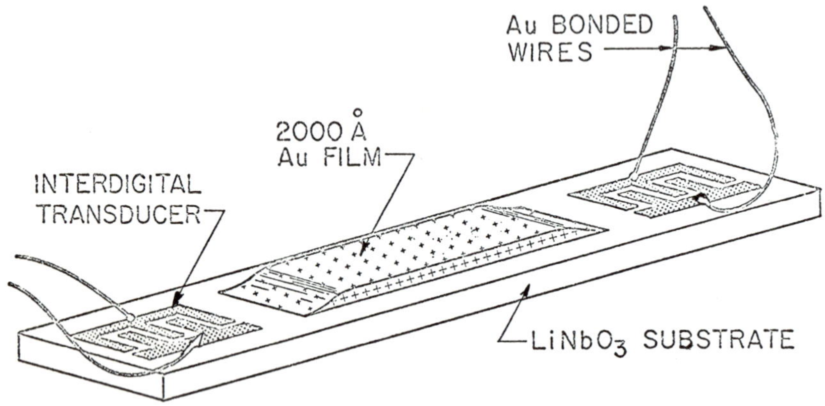

Fig. 4

AMPLIFIER CONFIGURATIONS

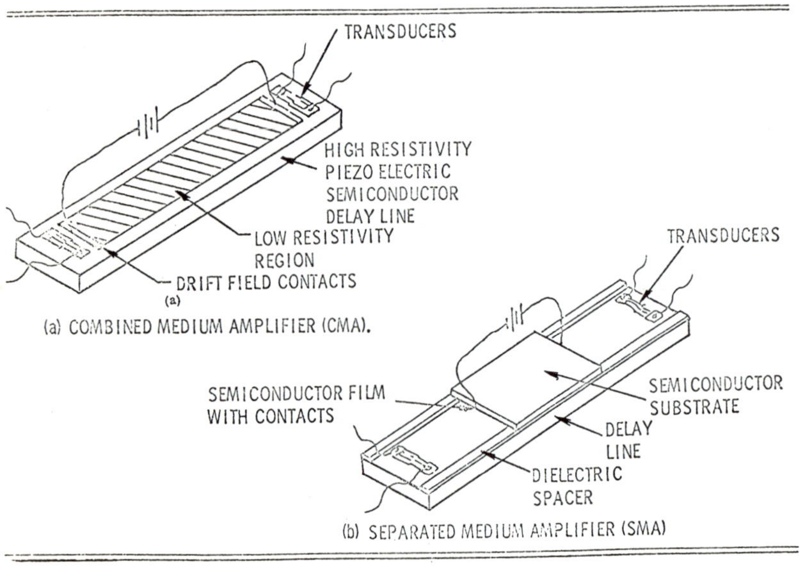

Fig. 5

A nonlinear surface elastic wave device which delivers the convolution of its input signals is[66,67,102] shown in Fig. 3. Such nonlinear devices can provide completely adaptive, analog matched filters, signal correlation, pulse compression, fourier transformation and electronically variable time delay. A typical dispersive delay line is illustrated in Fig. 4. A short input pulse will be expanded by such a device while an FM input signal can be time compressed.[104]

Typical surface elastic wave amplifiers[22,68,69,100,105,106] are shown in Fig. 5. In such amplifiers, drifting carriers in a semi-conductor adjacent to a piezoelectric crystal surface couple to the traveling electric field associated with an elastic surface wave propagating on the crystal surface. When the drift voltage is adjusted so that the carrier drift speed exceeds the Rayleigh wave speed, energy transfer to the Rayleigh wave takes place. Laboratory built silicon on sapphire Rayleigh wave amplifiers, operating at 108 MHz, have yielded terminal, or net, gains of 30 dB for d-c input power of only 0.7 watts.[100]

Conclusions

Linear surface elastic wave propagation and excitation are fairly well understood, although the exact details of operation of an interdigital electrode array embedded in or on the surface of an arbitrarily anisotropic semi-infinite layered piezoelectric medium have not been worked out yet.

A great deal of theoretical and experimental work remains to be done in understanding the physics of nonlinear surface elastic wave devices.[65-67] Information on the variation of elastic surface wave velocity with temperature for various crystal cuts and orientations is unavailable except for a few crystals. In the area of devices, octave bandwidth surface wave delay lines yielding hundreds of microseconds of delay are a design goal. So are pulse compression filters with pulse compression ratios of 5000:1 or better. Programmable tapped delay lines using silicon or sapphire diode arrays for tap switching integrated on a single zinc oxide covered sapphire substrate are currently under development. In addition to being able to act as matched filters for variable codes, such programmable tapped delay lines can be used in pseudo-random code generation schemes. Nonlinear surface elastic wave convolvers[66,67] are entering into the development stage and systems applications are anticipated in the near future. Two major drawbacks to elastic surface wave amplifiers are failure to achieve consistent CW operation and the requirement of a fraction of a wavelength airgap between the semi-conductor and the adjacent piezoelectric substrate. Monolithic structures eliminating the airgap have been successfully operated,[69,106] however CW operation is still a problem.

References

1. F. Press, Science 133, 1455 (1961)
2. E. G. Lean and A. N. Broers, Microwave Journal, 13, 97 (1970).
3. J. L. Synge, J. Math. Phys. 35, 323 (1957).
4. J. L. Synge, Proc. Roy. Irish. Acad. A58, 13 (1956).
5. G. W. Farnell, Properties of Elastic Surface Waves, in Physical Acoustics Vol. VI, Chap. 3, p. 109, Ed. by W. P. Mason and R. N. Thurston, Acad. Press, N.Y., 1970.
6. T. C. Lim and G. W. Farnell, J. Appl. Phys. 39, 4319 (1968).
7. T. C. Lim and G. W. Farnell, J. A.S.A., 45, 845 (1969).
8. R. G. Pratt and T. C. Lim, Appl. Phys. Lett., 15, 403 (1969).
9. B. R. Tittmann, E. A. Kraut, and T. C. Lim, Proc. IEEE. 59, 86 (1971).
10. J. J. Campbell and W. R. Jones, J. Appl. Phys. 41, 2796 (1970).
11. G. Koerber, IEEE Trans. on Sonics and Ultrasonics, SU-18, 73 (1971).
12. G. W. Farnell, IEEE. Trans. on Sonics and Ultrasonics, SU-17, 229 (1970).
13. J. O. Bleustein, Appl. Phys. Lett. 13, 412 (1968).
14. Y. V. Gulyaev, Sov. Phys. JETP. 9, 63 (1969).
15. I. A. Viktorov, Rayleigh and Lamb Waves, Physical Theory and Applications, Plenum Press, New York, 1967.
16. H. F. Tiersten, Linear Piezoelectric Plate Vibrations, Plenum Press, New York, 1969.
17. L. M. Brekhovskikh, Waves in Layered Media, Academic Press, New York, 1960.
18. W. M. Ewing, W. J. Jardetzky, and F. Press, Elastic Waves in Layered Media, McGraw-Hill, Book Co., New York, 1957.
19. J. E. White, Seismic Waves, Radiation, Transmission, and Attenuation, McGraw-Hill Book Co., New York, 1965.

20. R. Holland and E. P. Eernisse, Design of Resonant Piezoelectric Devices, Research Monograph No. 56, The MIT Press, Cambridge, Mass., 1969.

21. M. J. P. Musgrave, Crystal Acoustics, Holden-Day, San Francisco, 1970.

22. R. M. White, Proc. IEEE. 58, 1238 (1970).

23. K. Dransfeld and E. Salzmann, Excitation, Detection, and Attenuation of High Frequency Elastic Surface Waves, in Physical Acoustics Vol. VII, Chap. 4, p. 219, Ed. by W. P. Mason and R. N. Thurston, Acad. Press, N.Y., 1970.

24. J. de Klerk, Ultrasonics 9, 35 (1971).

25. H. Sabine and P. H. Cole, Ultrasonics 9, 103 (1971).

26. H. I. Smith, Intl. Jo. of Nondestructive Testing 2, 31 (1970).

27. E. A. Kraut, Rev. of Geophysics 1, 401 (1963).

28. A. B. Smith and R. W. Damon, IEEE. Trans. on Sonics and Ultrasonics SU-17, 86 (1970).

29. C. C. Tseng and R. M. White, J. Appl. Phys. 38, 4274 (1967).

30. Lord Rayleigh, Proc. Lond. Math. Soc. 17, 4 (1885).

31. R. Stoneley, Proc. Roy. Soc. A232, 447 (1955).

32. D. C. Gazis, R. Herman, and R. F. Wallis, Phys. Rev. 119, 533 (1960).

33. V. T. Buchwald and A. Davis, Quart. J. Mech. Appl. Math. 1b, 283 (1963).

34. D. A. Tursonov, Sov. Phys. - Acoust. 13, 78 (1967).

35. R. Burridge, Quart. J. Mech. and Appl. Math. 23, 217 (1970).

36. H. Lamb, Phil. Trans. Roy. Soc. (London) A203, 1 (1904).

37. E. A. Kraut, Propagation of a pulse from a surface line source on a transversely isotropic elastic half space, Ph.D. thesis, University of California, Los Angeles, 1962.

38. I. Abubakar, Geophys. J. 6, 337 (1962).

39. I. Abubakar, Geophys. J. 7, 87 (1962).

40. R. Burridge, Quart. J. Mech. and Appl. Math. 24, 81 (1971).

41. H. I. Smith, F. J. Backner, and N. Efremow, J. of Electro-Chem. Soc. 118, 821 (1971).

42. R. M. White, IEEE Trans. on Electron Devices ED14, 181 (1967).

43. C. C. Tseng, IEEE Trans. on Electron Devices ED15, 586 (1968).

44. H. Engan, IEEE Trans. on Electron Devices ED16, 1014 (1969).

45. S. G. Joshi, and R. M. White, J.A.S.A. 46, 17 (1969).

46. G. A. Coquin and H. F. Tiersten, J.A.S.A. 41, 921 (1967).

47. H. Skeie, J.A.S.A. 48, 1098 (1970).

48. N. Smith, H. Gerard, J. Collins, T. Reeder, and H. Shaw, IEEE Trans. on Micro Wave Theory and Tech., MTT-17, 856 (1969).

49. W. Smith, H. Gerard, J. Collins, T. Reeder, and H. Shaw, IEEE Trans. on Micro Wave Theory and Tech. MTT-17, 865 (1969).

50. M. O. Vassell and A. K. Ganguly, IEEE Trans. on Electron Devices ED-18, 66 (1971).

51. B. A. Auld and G. S. Kino, M. L. Report No. 1841, Microwave Laboratory, W. W. Hansen Laboratories of Physics, Stanford University, Mar. 1970.

52. J. J. Cambell and W. R. Jones, IEEE Trans. on Sonics and Ultrasonics SU-15, 209 (1968).

53. E. A. Kraut, B. R. Tittmann, J. J. Graham, and T. C. Lim, Appl. Phys. Lett. 17, 271 (1970).

54. A. J. Slobodnik, Jr. and E. O. Conway, Microwave Acoustic Handbook, Vol. 1 Airforce Cambridge Research Laboratories, L. G. Hanscom Field, Bedford Mass. (1970).

55. K. A. Ingebrigtsen, J. Appl. Phys. 40, 2681 (1969).

56. M. B. Schulz, B. J. Matsinger, and M. G. Holland, J. Appl. Phys. 41, 2755 (1970).

57. F. S. Welsh, IEEE Trans. On Sonics and Ultrasonics SU-18, 108 (1971)

58. R. G. Steg and P. G. Klemens, Phys. Rev. Lett. 24, 381 (1970).

59. J. A. Hudson, Proc. Camb. Phil. Soc. 67, 215 (1970).

60. P. H. Carr, IEEE Trans. on Microwave Theory and Tech. MTT-17, 845 (1969).

61. F. Rieschbieter, Acoustica 16, 75 (1965).

62. F. Rieschbieter, Acoustica 18, 109 (1967).

63. L. G. Lean, C. G. Powell and L. Kuhn, Appl. Phys. Lett. 15, 10 (1969).

64. E. G. Lean and C. C. Tseng, J. Appl. Phys. 41, 3912 (1970).

65. L. O. Svaasand, Appl. Phys. Lett. 15, 300 (1969).

66. M. Luukkala and G. S. Kino 18, 393 (1971).

67. E. A. Kraut, T. C. Lim and B. R. Tittmann, IEEE Symposium on Applications of Ferroelectricity, 1971, IEEE Transactions (to be published).

68. K. M. Lakin and H. J. Shaw, IEEE Transactions on Microwave Theory and Techniques MTT-17, 912 (1969).

69. L. A. Coldren and G. S. Kino, Appl. Phys. Lett. 18, 317 (1971).

70. W. Smith, T. M. Reeder, J. H. Collins, and H. J. Shaw, IEEE Transactions on Microwave Theory and Techniques MTT-17, 1043 (1969).

71. M. R. Daniel and J. de Klerk, Appl. Phys. Lett. 16, 331 (1970).

72. H. F. Tiersten, J. Appl. Phys. 40, 770 (1969).

73. L. R. Adkins and A. J. Hughes, J. Appl. Phys. 42, 1819 (1971).

74. J. H. Collins and P. J Hagon, Electronics 42, 102 (1969).

75. R. V. Schmidt and F. W. Voltmer, IEEE Transactions on Microwave Theory and Technique MTT-17, 920 (1969).

76. T. C. Lim, E. A. Kraut, and B. R. Tittmann, Appl. Phys. Lett. 17, 34 (1970).

77. K. Sezawa and K. Kanai, Bull. Earthquake Res. Inst. 13, 237 (1935).

78. I. Tolstoy and E. Usdin, Geophysics 18, 844 (1953).

79. A. E. H. Love, Some Problems in Geodynamics, Cambridge, University Press, 1911.

80. J. C. Scholte, Roy. Astr. Soc. Monthly Notices Geophys. Suppl. 5, 120 (1947).

81. R. Stoneley, Proc. Roy. Soc. A106, 416 (1924)

82. M. Newlands, Phil. Trans. Roy. Soc. (London) A245, 213 (195).

83. T. C. Lim and M. J. P. Musgrave, On Elastic Waves at Crystalline Interfaces, Mathematics Dept., Imperial College of Sci and Tech., Univ. of London (1969).

84. W. W. Johnson, Bull. Seism. Soc. of Am. 60, 1105 (1970).

85. F. Schwab, Bull. Seism. Soc. of Am. 60, 1491 (1970).

86. S. Crampin, Geophys. J. R. Astr. Soc. 21, 387 (1970).

87. L. P. Solie, Appl. Phys. Lett. 18, 111 (1971).

88. E. A. Kraut, Phys. Rev. 188, 1450 (1969).

89. F. Gilbert and G. Backus, Geophysics 31, 326 (1966).

90. F. Abramovici, Bull. Seism. Soc. Am. 58, 427 (1968).

91. F. Abramovici, Geophys. J. R. Astr. Soc. 16, 9 (1968).

92. K. A. Ingebrigtsen and A. Tonning, Phys. Rev. 184, 942 (1969).

93. K. A. Ingebrigtsen and A. Tonning, IEEE Trans. on Microwave Theory and Technique MTT-17, 827 (1969).

94. K. A. Ingebrigtsen, E.LAB Report TE-94, Electronics Res. Lab., Norwegian Inst. of Tech., Trondheim, Norway, Nov. 1967.

95. S. J. Laster, J. G. Foreman, and A. F. Linville, Geophysics 30, 571 (1965).

96. F. Gilbert, Reviews of Geophys. 2, 123 (1964).

97. C. L. Pekeris, Z. Alterman, F. Abramovici, and H. Jarosch, Rev. of Geophys. 3, 25 (1965).

98. Z. Alterman and F. C. Karal, Jr., Bull. Seism. Soc. of Am. 58, 367 (1968).

99. J. H. Collins and P. J. Hagon, Electronics 42, 94 (1969).

100. J. H. Collins and P. J. Hagon, Electronics 42, 102 (1969).

101. J. H. Collins and P. J. Hagon, Electronics 43, 110 (1970).

102. R. Gundlach, Electronics 43, 93 (1970).

103. R. H. Tancrell, Proc. IEEE 59, 393 (1971).

104. C. Lardat, C. Maerfeld, and P. Tournois, Proc. IEEE 59, 355 (1971).

105. H. Hanebrekke, K. A. Ingebrigtsen, Electronics Lett. 6, 16 (1970).

106. C. W. Turner, CRC. Critical Rev. in Solid State Sci., 4, 575 (1970).

GUIDED ACOUSTIC SURFACE WAVES: PRINCIPLES AND DEVICES*

L. R. Adkins and A. J. Hughes
Autonetics Division of North American Rockwell Corporation
Anaheim, California 92803

I. Introduction

Surface acoustic waves may provide the key to miniature signal processing systems in the future. The fact that these waves can be tapped, guided, spatially processed, or otherwise manipulated on a miniature substrate makes the technology ideally suited to a range of processing components.

Thus far, devices such as surface-acoustic-wave (SAW) delay lines, pulse compression filters, and frequency filters have received the bulk of the research and development attention. These devices have, in general, employed line-of-sight acoustic wave propagation and thus are not guided-wave devices. These devices therefore do not begin to exhaust the full potential which may ultimately be realized with guided waves and a sophisticated SAW technology integrated with microelectronic semiconductor technology.

The present authors have been engaged in a combined theoretical and experimental research effort on surface acoustic wave guidance, primarily the slow-on-fast film-substrate combination, over the past two years, and this paper is a survey of the significant results of that program. The subsequent discussion will deal with surface acoustic wave guides and their applications to couplers and delay lines. For discussions on other aspects of surface wave acoustics the reader is referred to articles and references in this volume, and for a compact survey of surface wave acoustic work up to late 1969, to the Microwave Theory and Techniques special issue devoted entirely to microwave acoustics (Ref. 1). Also to be noted is a brief summary article by Worley (Ref. 2) on acoustic surface waves as a practical microwave technology. In addition, an investigation of a type of guided wave device not discussed in the present paper - the ring resonator - has been reported by Owen and Knox (Ref. 3).

Surface-acoustic-wave guidance is of fairly recent origin. Ash (Ref. 4) and Ash and Morgan (Ref. 5) demonstrated wave guidance by a pair of grooves inscribed in an elastic substrate surface. This guidance structure is called topographic, in that guidance is due to topographical surface deformations such as grooves, slots, or ridges on the substrate. Such guides have been more recently reviewed by Ash et al (Ref. 6); ridge-type waveguides have been explored theoretically by Waldron (Ref. 7).

Although an adequate characterization of topographic guidance structures cannot presently be made, some potential difficulties of this approach have emerged. The fabrication of topographic guides at the frequencies of most interest appears rather difficult and may impose severe limitations on this guidance technique.

Almost coincident in time with the topographic studies, White (Ref. 8) reported investigations of a different wave guidance structure. Energy was channeled and wave guidance achieved by depositing a thin strip of material of acoustically slower-velocity on a substrate of a material of acoustically faster-velocity. White reported experimental results for gold strips deposited on fused silica. Stern (Ref. 9) has described this guidance structure as the acoustic analog of the electromagnetic dielectric waveguide. Physically, the acoustic wave is evanescent in the faster-velocity region of the substrate and thus becomes effectively bound to the slow-velocity region.

The pioneering theoretical investigation of guidance structures which depend upon a velocity differential is that of Tiersten (Ref. 10), who treated both strip (slow-on-fast) and slot (fast-on-slow) guidance structures. Numerical dispersion curves were presented for gold on fused silica and Al on T-40 glass. The calculated velocities were in good agreement with White's experimental data. Tiersten also developed a weak coupling model for double guide structures.

The gold-on-quartz strip-guide structure has been investigated extensively at Autonetics by the present authors (Ref. 11), with both theoretical and experimental studies involved. In Section II we will briefly outline our theoretical model for a single gold film on fused quartz guide and present experimental results which substantiate this model. The model predicts velocity, transverse decay lengths and other guidance parameters.

The guidance model developed by Adkins and Hughes has also been extended to the case of two parallel guides sufficiently close together to interact acoustically (Refs. 12, 13, 14) and is described in Section III along with the measurements performed on these structures. The acoustic interaction between guides was treated exactly and the coupling lengths calculated are valid for all coupling strengths.

The experimental studies and the theoretical model developed for SAW guidance have proven adequate for development of experimental directional couplers, Y-junction couplers and long folded-path delay lines (Refs. 13, 15, 16). Results of the investigations of these initial guided wave devices are presented in Section IV. Although the materials combination employed in these studies was gold on fused and crystalline quartz, the general features of the devices have been established and can be applied to other materials combinations.

One feature which has emerged from the device work is that the surface-wave propagation is sufficiently dispersive to rule out many practical device applications. The basic physics of a guidance structure which is substantially non-dispersive has been explored theoretically (Ref. 17), and is briefly described in Section V. These theoretical results appear promising and suggest that the proposed method of guidance can play an important role in the growing surface-wave technology.

The paper is concluded in Section VI with a brief summary and discussion. The importance of heteroepitaxial materials efforts for realization of a sophisticated acoustic technology is stressed.

II. Single Gold-on-Fused Quartz Guides

In this section we shall briefly summarize our theoretical and experimental studies (Ref. 11) of single gold-on-quartz guides and will present the main elements of the theoretical guidance model which we have developed for isotropic guide/substrates.

The general theoretical approach to the slow film on fast substrate (strip-guide) problem was pioneered by Tiersten (Ref. 10). Although our model is based upon that of Tiersten, there are significant differences involved in the use of different acoustic variable constraints for solution.

Figure 1 shows an end view of a gold strip guide (width 2d and thickness H) deposited on an isotropic elastic substrate. The direction of guided wave propagation is out of the paper and is the X_1 direction. The surface wave amplitude dies off exponentially in the X_2 direction. The acoustic signal amplitude is transversely localized in the X_3 direction and is assumed to be $\approx \cos \bar{k}_3 X_3$ in Region I of Fig. 1 and to decay exponentially $\approx \exp - k_3(|X_3| - d)$ in Regions II. Here $\bar{k}_3$ and k_3 are transverse k-vectors in Regions I and II, respectively.

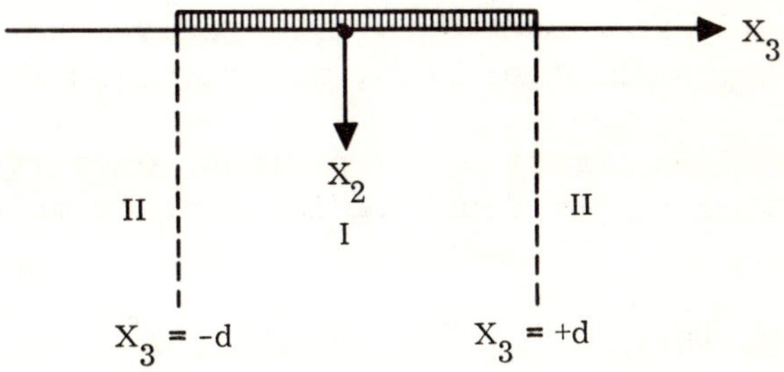

Fig. 1 End view of gold guide on quartz substrate

Our solution consists essentially of matching velocities of bound mode propagation in Regions I and II and of matching acoustic field variables ψ_3 and ψ_1 across the boundaries $X_3 = \pm d$ between the two regions. Consider the velocity constant first. Because of the slow acoustic velocity gold film in Region I, the ordinary (straight-crested and of uniform amplitude in the ψ_3 direction) Rayleigh wave velocity is slower in Region I than in Region II.

To secure the velocity match required for spatial coherence, we employ cosine-cresting of the plane wave in Region I to increase the velocity there and employ exponential-cresting in Region II to decrease the velocity there (Refs. 10, 11). Sufficient increase and decrease, respectively, yields an acoustic velocity match.

We now consider constraints on the acoustic field variables employed. For an isotropic elastic problem and for Rayleigh waves, it is convenient to work with vector (ψ) and scalar (φ) potentials rather than with elastic displacements. We shall, however, need to consider only two variables ψ_3 and ψ_1 of the above four variables (ψ_1, ψ_2, ψ_3, φ). For solution, we require continuity of ψ_3 and ψ_1 across the boundary $X_3 = \pm d$. Tiersten, on the other hand, required continuity of ψ_3 and the normal derivative of ψ_3 for a solution. A unique solution is now possible in either case.

The mathematical solutions yield the bound mode velocity (V/β), and the transverse k-vector $\bar{k}_3$ (Region I) and k_3 (Region II); all in terms of the guide parameter $G = d/H$ and the independent variables k_1, H or ωH. Here k_1 is the bound mode propagation direction k-vector and ω is the frequency $2\pi f$.

The physical observables which can be conveniently measured experimentally are the velocity (V/β) and the decay length (X_d/λ). The decay length (X_d/λ) is, in units of wavelength, the distance away from the guide edge required for the acoustic amplitude to decay to 1/e. In terms of k_3, $(X_d/\lambda) = k_3/(k_1 2\pi)$.

The velocity dispersion and the decay lengths have been measured experimentally for several gold guides. These experiments were performed at frequencies sufficiently low (5 MHz to 14 MHz) to enable us to prepare surface wave apparatus of maximum versatility. The surface waves were launched and detected by piezoelectric PZT4 transducers mounted on polystyrene wedges. Experimental data points for the velocity as a function of frequency were plotted for the guides investigated, along with the theoretical dispersion curves. Agreement between experiment and theory is substantial, and within experimental error. Decay lengths also agree with theory and these, along with other data, are given in Ref. 11.

The good agreement between theory and experiment encouraged us to extend this approach to the solution of the double guide problem which is discussed in the next section.

III. Double Guide Structure

We now consider the case of a gold on fused quartz guidance structure comprised of two parallel guides spaced closely enough together to produce acoustic coupling. The key phenomenon here is that an acoustic signal launched on one guide of a double guide system will oscillate back and forth between the two guides as the energy propagates down the structure. The coupling length is defined as the distance in the propagation direction required for an energy distribution on one line to transfer to the other line. The calculation of this coupling length, along with other physical observables of the mode coupling phenomenon, will, therefore, be emphasized in the development of a double guide theoretical model. This model forms the basis for the coupled mode directional coupler discussed in the next section.

The mathematical description of the double guide structure is in terms of the two lowest order normal modes--a symmetric double guide mode and an antisymmetric double guide mode. Because of physical symmetry between the two guides, each of these modes must, individually, produce an equal amount of acoustic power on each of the two guides.

An acoustic signal preferentially launched on one of the guides generates a linear combination of antisymmetric (modal amplitude A) and symmetric (modal amplitude B) double guide modes. The two modes add constructively on one guide and destructively on the other guide and produce a resultant mode which is a maximum on the launch guide. The two basic normal modes propagate down the double guide structure with slightly different velocities and thus change in relative phase.

Suppose the modes are initially "in phase" and a maximum on, say the left hand guide. Then at a propagation distance L down the structure the modes are "out of phase" and the maximum is now located on the right hand guide. This process then repeats itself down the structure to the end of the guides. The distance L is the coupling length referred to above and is calculated from the difference in propagation k-vectors of the two modes.

The first theoretical calculation of coupling length for a double guide structure was that of Tiersten (Ref. 10) who developed a theoretical model using perturbation techniques for coupling based on the assumption of weak coupling between the guides. No experimental measurements were then available for comparison with the weak coupling theory.

We have developed a model (Refs. 12, 13, 14) which exactly treats the interaction between guides and is thus valid, within the framework of the model, for arbitrary coupling strengths.

Briefly, following a modification of the approach described for single guides in Section II, double guide modes are constructed from solutions having exponential cresting in the region outside the double guide structure, harmonic cresting on the guides, and hyperbolic cresting between the guides.

This model has allowed us to calculate coupling lengths L, velocity dispersion relations (V/β), decay lengths (X_d/λ), and antisymmetric/symmetric modal amplitude ratios (A/B) for any isotropic double guide structure with given geometrical parameters guide width 2d, film thickness H, and guide separation 2c. This theoretical model has been substantiated through measurements of guidance structure observables.

Measurements of both individual mode and resultant mode guidance properties were performed. Decay length measurements were made and found to be in good agreement with calculated decay lengths for the two basic modes.

The primary resultant mode property measured was the coupling length L. Two features of possible importance in practical device applications are predicted by our exact coupling model but do not appear in the weak coupling approximation, namely a low frequency cutoff below which coupling does not take place and a coupling length local minimum located in the strong coupling strength frequency domain.

The significance of the low frequency cutoff for coupling phenomenon is clear. The coupling length minimum may prove important in device design, since the coupling length is approximately linear with respect to frequency about this minimum. Thus wider bandwidth devices may be possible here than for weak coupling devices. Comparison with our results indicates that the weak coupling theory yields coupling lengths which are numerically too small.

Other resultant mode properties of practical interest were also measured. The modal amplitude ratio (A/B) is the relative amount of antisymmetric and symmetric modes present and is a function of frequency and of the geometrical parameters. Measurements of (A/B) were found to be in reasonable agreement with calculated ratios. The modal amplitude ratio is important in determining the attainable power ratio between guides since this ratio depends upon constructive and destructive interference between modes. Optimum cancellation will, of course, occur at high frequencies for which (A/B) = 1. Typical power ratios ranged from 4 dB near cutoff to more than 20 dB at high frequencies.

Measurements of the above guidance parameters were made on a number of different gold-on-quartz double guide structure. The features predicted theoretically by the model--the behavior of decay lengths, modal amplitude ratios and coupling lengths--were confirmed unambiguously and good numerical agreement was obtained. For example, the agreement for the coupling lengths is within experimental error (10-15%). Since the coupling length is a strong function of geometrical parameters and is, above the local minimum, approximately an exponential function of frequency, agreement here represents a particularly severe test of both the theoretical model developed and the experimental approach employed.

The information obtained from the basic studies discussed in this and the preceding section allowed us to design and fabricate several prototype devices which are described below.

IV. Guided Surface Acoustic Wave Devices

Three different prototype guidance devices have been fabricated and tested. These devices are the coupled mode directional coupler (Refs. 13, 15), the so-called Y-junction coupler (Ref. 15) and the folded-path long delay line (Ref. 16). All of these devices employed gold on quartz guidance structures. The coupled mode directional coupler and the folded-path delay devices were investigated by means of wedge transducers and were on fused quartz substrates. The Y-junction couplers employed interdigital transducers and were on crystalline quartz. Design parameters for guidance structures on crystalline quartz were estimated from the fused quartz data and the isotropic guidance model.

The coupled mode directional coupler was based on the parallel guide structure discussed in Section III. Briefly, two guides, one longer than the other, are deposited on fused quartz and are drawn apart after a parallel interaction region equal to a predetermined fraction of one or more coupling lengths at a desired operating frequency. This configuration is shown in Fig. 2.

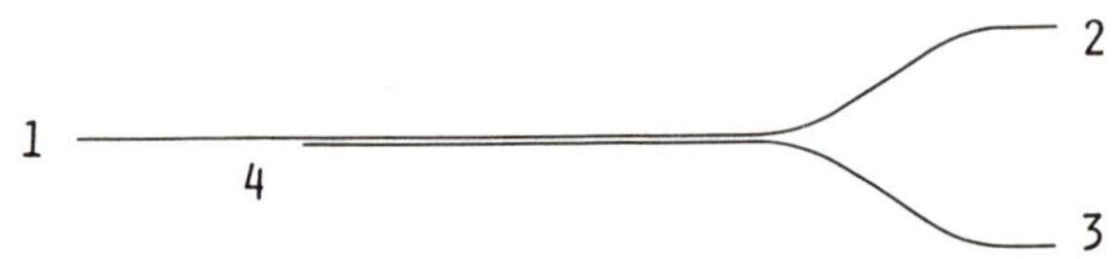

Fig. 2 Coupled mode double guide coupler

In practice, coupled mode directional couplers would consist of two pairs of symmetric ports. In the device investigations reported here, our goal was to unambiguously correlate the coupler output arm amplitudes with coupling lengths measured in the parallel portion of the device. It was therefore decided to employ the input configuration investigated in the parallel double guide studies. Conditions required for unambiguous generation of pure double guide modes and a precisely located spatial initiation of coupling were well understood. We have therefore designed and fabricated the structure as shown in Fig. 2. However, subsequent device investigations on this structure indicate that the design (length, radius of curvature and separation angle) of the exit arms for output ports 2 and 3 was adequate and could easily have been employed at ports 1 and 4.

Let us launch an acoustic signal at port 1 in Fig. 2 and consider the output amplitudes at ports 2 and 3. A crucial question for device operation in terms of a coupling length model is whether the acoustic energy distribution between the guides at the point of divergence (end of the parallel portion of the double guide structure) holds as the guides are separated. We have found experimentally that this is the case (Refs. 13, 15).

A coupler with appropriate geometrical design parameters was fabricated and acoustic amplitude ratios at the output arms were measured as a function of frequency from about 4.7 MHz to 9.0 MHz. The surface waves were launched and detected by means of wedge transducers on micropositioners. The amplitude ratio (A_3/A_2) at ports 3 and 2 was found to be a maximum at frequencies for which the length of the parallel portion of the device was an odd integral number of coupling lengths (in this case 1 and 3), a minimum at an even integral number of coupling lengths (here 2) and equal to one at odd multiples of half a coupling length (0.5, 1.5, 2.5, and 3.5). An additional minimum in the ratio (A_3/A_2) was found, corresponding to the local minimum in the coupling length versus frequency curve in the strong coupling region discussed in Section III.

Thus, in principal, predetermined fractions of acoustic energy on a waveguide operating at a given frequency may be coupled out by designing a double guide structure with the proper coupling length. However, for any practical device additional characteristics--output arm isolation, bandwidth, and losses--must also be considered.

The isolation between arms 2 and 3 is defined as the ratio between the power input at arm 2 and the power output at arm 3. This ratio was measured as a function of frequency for the coupler discussed above and found to be between 28 and 32 dB for the frequency range from 5.0 to 8.5 MHz. In making these isolation measurements, black wax was employed at ports 1 and 4 for resistive terminations.

Coupler loss information was obtained by comparing the sum of the powers in the output arms (ports 2 and 3) to the input power (port 1). These measurements were made as a function of frequency from 5 MHz to 9 MHz. Our data was taken with free standing wedges and exhibited scatter. Roughly, there was an indicated loss of about 5.5 dB at the low frequency end, a minimum of about 1 dB around 6 MHz and an increase to about 4 dB at 9 MHz.

The bandwidth of this device is quite narrow by conventional microwave standards. For example, 3 dB microwave couplers often have a device bandwidth defined in terms of 3 dB $\pm$ 0.15 dB. On this basis the (A_3/A_2) data discussed above indicate that the subject device has about 0.6% bandwidth. This, however, is not an optimum case, since for experimental purposes we have selected a parallel region containing several coupling lengths. It can be shown that if we shorten this region to contain one-half coupling length at 6.2 MHz, the 3 dB $\pm$ 0.15 dB bandwidth would be about 5%. This is still quite narrow.

To summarize the investigation of this device we point out that although adequate coupling and loss characteristics are attainable with a coupled mode structure, the narrow bandwidth limits its range of application as a directional coupler. However, it is possible that this narrow bandwidth characteristic may be employed to good advantage in the design of other signal processing devices such as acoustic filters and discriminators.

The second device to be discussed, shown in Fig. 3, we have labeled the Y-junction coupler. Y-junction couplers operating at 25 MHz and 100 MHz have been fabricated and investigated in detail.

Fig. 3 Y-junction coupler

There are two major features of this design. First, the device is a non-coupled mode structure in that a single mode is generated by an acoustic signal input at port 1. Second, the device has physical symmetry about a center line. Thus, for a given input at port 1 equal powers must be delivered to ports 2 and 3. The device is an inherently broadband 3 dB coupler for an input at port 1. For inputs at ports 2 or 3, device operation does, however, involve mode coupling and must be viewed as a four port device as shown in Fig. 3.

Matching horns have been provided at each of the three arms as transitions between the wide (25 wavelengths) interdigital transducers and the narrow ($\approx$one wavelength) guidance structure. Measurements indicated that each horn introduced a loss of approximately 5 dB. The preliminary nature of these measurements suggests that a more thorough investigation of such structures would be desirable.

The complete system (transducers, horns and couplers) was characterized for insertion loss, output arm isolation, and bandwidth. At 26.1 MHz the total insertion loss was found to be 36 dB with identical results being obtained for ports 1 $\rightarrow$ 2 and 1 $\rightarrow$ 3. To measure the system isolation, signal was fed into port 2 and received at ports 1 and 3. The insertion loss 2 $\rightarrow$ 1 was found to be $\approx$36 dB, which is identical to the 1 $\rightarrow$ 2 value although in the former case a 3 dB power split between output arms was present. For an input at port 2, the isolation of port 3 was between 40 dB (72μsec delayed reflection from port 1) and 50 dB (36μsec delayed reflection from the Y-junction). In all measurements reported near 25 MHz, the electric feed-through signal was >70 dB down from input levels.

The measured 3 dB bandwidth of the device was found to be 5.5%. This measured bandwidth is determined by the characteristics of the transducers and tuning elements and does not represent the bandwidth of the coupler structure which itself is broadband.

A Y-junction coupler operating near 100 MHz was also investigated. Substantially identical insertion loss, isolation and bandwidth results were obtained as for the preceding 25 MHz device. The slight differences in characteristics apparently could be attributed to tuning and stray capacitance of the transducers. Other factors, such as increased propagation losses in the gold at higher frequencies, will play an important role in characteristics of some guided wave devices but could not be resolved in the present experiments.

To conclude this discussion of the Y-junction coupler let us consider two aspects of the intrinsic acoustic characteristics of the device. First consider the intrinsic acoustic loss. With optimum tuning (which narrows the bandwidth) a total insertion loss (ports 1 → 2) of 32 dB can be obtained at 25 MHz. From this figure one subtracts 3 dB for the equal arm power split, 16 dB for the transducers (independently measured), and about 10 dB for two horns. Thus, we are left with a maximum intrinsic Y-junction coupler loss of about 3 dB.

The second aspect concerns a basic characteristic of this type of structure. This involves the interpretation of the combined insertion loss and isolation measurements. Note that the insertion loss for 1 → 2 was identical with that for 2 → 1. The input-output path 1 → 2 involved a 3 dB power split into port 3 while the path 2 → 1 involved no such split since port 3 was isolated by 50 dB. This latter observation cannot be described by network theory for a lossless reciprocal three-port device with symmetry between two of the ports.

A pseudo-four port network model was developed in (Ref. 15) which predicts the above characteristics. The consideration of the fourth port is required by the fact that an input at port 2 or 3 generates both a symmetric and an antisymmetric waveguide mode. Thus port 4 relates to an antisymmetric (higher order) waveguide mode.

Summarizing, Y-junction couplers operating near 25 MHz and 100 MHz have been fabricated and loss and isolation measurements have been made. Very good isolation (better than 50 dB for the intrinsic coupler at 25 MHz) has been obtained and the apparent intrinsic coupler loss is low (less than 3 dB). The device is inherently broadband. Further improvements will require concentration on tuning and on horn and transducer design.

The final device to be discussed is the folded-path guided surface wave delay line. This structure offers a significant potential advantage over single-propagation path delay lines for obtaining very long delay times. For SAW delay lines the delay time is determined by the propagation path length which can be obtained on available substrates. Excluding wrap-around devices of the type discussed in Ref. 18, conventional device delay times are currently limited to about 100μsec. An acoustic waveguide folded into a spiral or serpentine configuration can increase the propagation path length, and hence the delay time, on a given substrate surface by a factor of perhaps 5-20. Of course, in order to utilize this greatly increased path length in practical devices, losses, spurious signal levels, and dispersion must be within acceptable limits.

We have designed and fabricated a spiral path guided SAW delay line and have measured the delay time, dispersion, loss, and spurious signal level. This device, shown in Fig. 4, employs gold on fused quartz acoustic waveguides, and the surface waves are launched and detected with wedge transducers. All measurements were made as a function of frequency covering a range from 6 to 21 MHz.

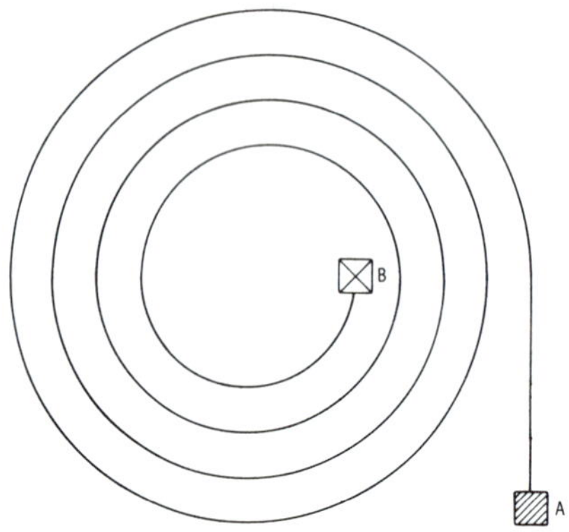

Fig. 4 Folded-path long delay line

The total delay time was measured by launching the surface waves at point A (Fig. 3) and detecting at point B. The generator was operated in the pulse mode so that the delay time could be read directly off an oscilloscope. The measured total time delay ranged from about 210μsec at 10 MHz to 260μsec at 21 MHz. The slope of an approximate straight-line fit to experimental data yields a dispersion of about 4.2μsec/MHz. This compares with theoretical estimates of about 5.1μsec/MHz.

The spurious signal level (i.e., signals with delay times different from the main guided signal) was measured with the transmit and receive transducers in the same on-guide position as in the delay time measurements. It was found that at frequencies above about 10 MHz the spurious signal level was very low, greater than 35 dB below the guided spiral path signal. At frequencies below 10 MHz the spurious signal increased sharply and below 7 MHz no guided spiral path signal was observable. The rapid increase of the spurious signal below a certain minimum frequency, here 10 MHz, serves to establish the operating frequency limit for a folded-path device with a given gold width, thickness and spacing between turns. The appropriate design parameters for a device with a desired operating frequency may be determined from the theoretical work on single and double guide structures discussed in Sections II and III.

The time differential loss of the device, which excludes transducer loss, was determined as a function of frequency by comparing the signal received at point B to the input signal on the guide at point A. There was considerable scatter in this data, but the average value of the loss in the frequency region from 10 MHz to 20 MHz was 0.036 dB/μsec. The highest loss measured (at 10 MHz) was 0.043 dB/μsec.

Summarizing, the delay time, spurious signal level, dispersion and propagation loss have been measured for a delay line employing a spiral path surface acoustic waveguide. The results are clearly encouraging in that spurious signal level and propagation loss are well within acceptable limits at these frequencies and are significant in that they establish that long delay times can be realized with guidance techniques.

The main potential use of the existing structure may lie in those device applications in which a significant differential time delay is required. On the other hand, the measured dispersion indicates that the existing structure would not be useful for non-dispersive applications since no adequate technique exists for compensating this much dispersion in practical devices. A guidance technique is briefly discussed in the next section which is substantially non-dispersive and may allow application of folded-path principles to non-dispersive devices.

V. Low Dispersion ($\Delta v/v$) Guidance Structure

Surface wave propagation in the structures discussed above is sufficiently dispersive to rule out many practical device applications. The basic physics of a guidance structure which is substantially non-dispersive, has been examined theoretically (Ref. 17). This guidance model employs the decrease Δv in surface-wave velocity v due to a thin conducting strip on a piezoelectric substrate to obtain a slow-velocity propagation path.

The ($\Delta v/v$) guidance structure should be intrinsically non-dispersive with the only dispersion present being that which arises solely from transverse confinement. This is small in comparison to the dispersion for gold strip guides which involve an elastic layer and are intrinsically dispersive. Typical theoretical examples of ($\Delta v/v$) guides on $LiNbO_3$ were from 300 to 1000 times less dispersive than gold strip guides on quartz.

Results of the theoretical model indicate that strong guidance and low dispersion are predicted for a number of materials in use or under development for microwave acoustics. This assumes that very thin conducting strips of the appropriate width are employed to define the waveguides. This required width is wider than in the case of the gold guides and is a strong function of the coupling strength of the particular material.

Bulk materials which could, theoretically, be employed include, but are not limited to, $LiNbO_3$, $Bi_{12}GeO_{20}$, and ZnO. Considerations of compatibility of ($\Delta v/v$) guidance with an integrated semiconductor-acoustic technology argue that heteroepitaxially deposited piezoelectrics on substrates such as sapphire (Al_2O_3) would be of major importance. On this basis AlN/Al_2O_3 and/or ZnO/Al_2O_3 look attractive for ($\Delta v/v$) guidance, based on materials growth and acoustic characterizations now underway at Autonetics.

VI. Summary and Conclusions

A theoretical model for velocity differential surface acoustic wave guides, which permits the calculation of device design parameters, has been developed and substantiated experimentally. The theoretical calculations have been employed in the design and fabrication of three prototype surface acoustic waveguide devices--the coupled mode coupler, the Y-junction coupler, and the folded-path delay line.

These devices have been tested and their pertinent operating characteristics determined. The present studies indicate that surface acoustic waveguide devices are feasible and that performance characteristics can be determined in a predictable manner. In addition, theoretical investigation of a different type of waveguide, the $(\Delta v/v)$ guide, has been discussed. This guide promises advantages over the previous type, especially with regard to dispersive characteristics, and warrants further investigation.

We feel that the present work represents a significant step toward the realization of an integrated acoustic surface wave technology, principally because it demonstrates that signal processing operations such as routing, splitting and combining of surface waves can be accomplished by means of suitably designed thin film structures deposited on fused or crystalline substrates.

The devices investigated have been individual discrete components. The potential usefulness of such devices in this stand-alone form is probably not great. For example, the couplers would have to compete with standard electromagnetic components, and the folded-path delay line would have to compete with helical "wrap-around" delay lines with non-dispersive long delay times.

The real systems impact of waveguide devices will lie in the development of an integrated surface acoustic wave technology with the realization of monolithic circuits in which surface acoustic waves will be interfaced with semiconductor large scale integrated technology. A two-fold effort will be required to bring this to fruition.

First, development of a wider class of surface acoustic guided wave components is required. A number of the basic building block guided wave devices have not yet been realized. The $(\Delta v/v)$ guidance structure appears to be a promising approach to devices requiring low dispersion. Also, continued progress toward a practical semiconductor-piezoelectric composite media amplifier is of major importance.

Second, assuming that a wider class of SAW devices can be developed and basic characteristics determined, the thrust toward a planar integrated practical technology will depend heavily on the continued development of hetero-epitaxial piezoelectric films, on substrate materials such as sapphire, suitable for acoustic devices operating a high frequencies and compatible with semiconductor processing and technology. In this sense acoustic materials investigations and development lie at the base of progress in this area.

REFERENCES

1. IEEE Trans. Microwave Theory Tech. MTT-17, No. 11 (Nov. 1969)
2. J. C. Worley, "Acoustic Surface Waves as a Practical Microwave Technology," Proceedings of the National Electronics Conference, Vol. XXVI, The Conrad Hilton Hotel, Chicago, IL, December 7-9, 1970 (paper 30A, p. 332)
3. D. B. Owen and R. M. Knox, "Waveguides and Distributed Components for Elastic Surface Waves," Proceedings of the National Electronics Conference, Vol. XXVI, the Conrad Hilton Hotel, Chicago, IL, December 7-9, 1970 (paper 38E, p. 702)
4. E. A. Ash, 1967 G-MTT Symp. Proc. (Boston, MA. May 9, 1967), p. 194
5. E. A. Ash and D. Morgan, Electronics Lett. 3, p. 462 (Oct. 1967)
6. E. A. Ash, R. M. De La Rue, and R. F. Humphryes, "Microsound Surface Waveguides," IEEE Trans. Microwave Theory Tech. MTT-17, pp. 882-892 (1969)
7. R. A. Waldron, "Mode Spectrum of a Microsound Waveguide Consisting of an Isotropic Rectangular Overlay in a Perfectly Rigid Substrate," IEEE Trans. Sonics and Ultrasonics SU-18, No. 1, p. 8 (January 1971)
8. D. L. White, 1967 IEEE Symposium on Sonics and Ultrasonics, Vancouver B.C., Canada. See Ref. 10 for a graph of White's experimental guided wave velocity data.
9. E. Stern, "Microsound Components, Circuits and Applications," IEEE Trans. Microwave Theory Tech. MTT-17, pp. 835-844 (1969)
10. H. F. Tiersten, "Elastic Surface Waves Guided by Thin Films," J. Appl. Phys. 40, pp. 779-789 (1969)
11. L. R. Adkins and A. J. Hughes, "Elastic Surface Waves Guided by Thin Films: Gold on Fused Quartz," IEEE Trans. Microwave Theory Tech. MTT-17, pp. 904-911 (1969)
12. L. R. Adkins and A. J. Hughes, "Acoustic Surface Wave Directional Coupler," 1969 IEEE Symp. on Sonics and Ultrasonics (St. Louis, MO, Oct. 23-24, 1969 (paper G-4)
13. L. R. Adkins, T. W. Bristol, P. J. Hagon and A. J. Hughes, "Surface Acoustic Waves - Device Applications and Signal Routing Techniques for VHF and UHF," Microwave Journal 13, No. 3, pp. 87-94 (1970)
14. A. J. Hughes and L. R. Adkins, "Coupling in Surface Acoustic Wave Guidance Structures," (unpublished June 1971)
15. L. R. Adkins and A. J. Hughes, "Investigation of Surface Acoustic Wave Directional Couplers," (unpublished February 1971)
16. L. R. Adkins and A. J. Hughes, "Long Delay Lines Employing Surface Acoustic Wave Guidance," J. Appl. Phys. 42, p. 1819 (April 1971)
17. A. J. Hughes, "Elastic Surface Wave Guidance by ($\Delta v/v$) Effect Guidance Structures," (unpublished June 1971)
18. W. L. Bond, T. M. Reeder, and H. J. Shaw, "Wrap-Around Surface-Wave Delay Lines," Electronics Lett. 7, No. 3, pp. 79 (February 1971)

---

*The research described in Section III and the acoustic coupler investigations in Section IV were supported in part by the U.S. Army Electronics Command under Contract DAAB07-69-0321.

PIEZOELECTRIC FILM SURFACE WAVE TRANSDUCERS

Dr. Fred S. Hickernell
Government Electronics Division
Motorola Inc.
Scottsdale, Arizona 85257

INTRODUCTION

A piezoelectric film layer in proximity to a periodic interdigital metal electrode can provide a simple and efficient means for the excitation and detection of surface elastic waves on nonpiezoelectric insulating, semiconducting, conducting, and amorphous substrates. The planar transducer configuration can be fabricated by standard semiconductor processing techniques. Using piezoelectric film transducer structures, surface elastic wave devices can be developed as an integral part of hybrid and monolithic microelectronic circuit assemblies.

Some of the transducer and device advantages of using a piezoelectric film transducer are the following: High fundamental frequency operation can be enhanced by the choice of a high velocity, low propagation loss material and by using a single phase array electrode geometry. Crystallographic orientation of the piezoelectric film permits the excitation of different surface wave propagation modes on layered media. Dispersive transducer and device structures are easily developed. Amorphous materials and crystals with low surface wave velocity anisotropy provide substrates for the development of surface elastic waveguides and associated waveguide components. Devices can be developed on substrates which enhance interactions between elastic waves and electrons, spin waves, optical waves and other elastic waves.

This paper describes the results of experimental work with piezoelectric film surface wave transducers. Zinc oxide (ZnO) and cadmium sulfide (CdS) films have been used to generate and detect surface elastic waves on a variety of nonpiezoelectric substrates with fundamental frequencies from 10 MHz to above 250 MHz. The CdS work has been described elsewhere.[1-3] This paper will primarily consider the work with ZnO films on fused quartz using a low frequency interdigital pattern. The data presented is illustrative of techniques the author has used to experimentally define the performance capabilities of different film and substrate combinations and develop design criteria for fabricating piezoelectric film transducers and associated surface wave devices.

BACKGROUND

In 1965 R. M. White and F. W. Voltmer[4] demonstrated that surface elastic waves could be generated efficiently on piezoelectric substrates with an interdigital electrode structure and suggested this technique could be extended to nonpiezoelectric substrates using piezoelectric films. The piezoelectric film technique has primarily been used to date for investigating the properties of Rayleigh and Love waves in layered media[5-9] and tapping surface wave energy.[10] It represents one of a group of methods which can be used for the excitation of surface waves on nonpiezoelectric substrates.[11] Recent theoretical work by Solie,[12] Smith,[13] Kino and coworkers[14] has helped to define the capabilities of piezoelectric film surface wave transducers and stirred renewed interest in this technology.

The four basic types of film transducer structures which may be considered for surface wave generation on nonpiezoelectric substrates are shown in Figure 1. They differ basically with respect to where the interdigital electrode and a conducting plane occurs with respect to the film and substrate. The position of the conducting surfaces influences both the electrical capacitance and the coupling efficiency of the device. The equivalent electrical circuit parameters for such configurations can be represented to good approximation by the basic circuit model developed by Smith and coworkers[15].

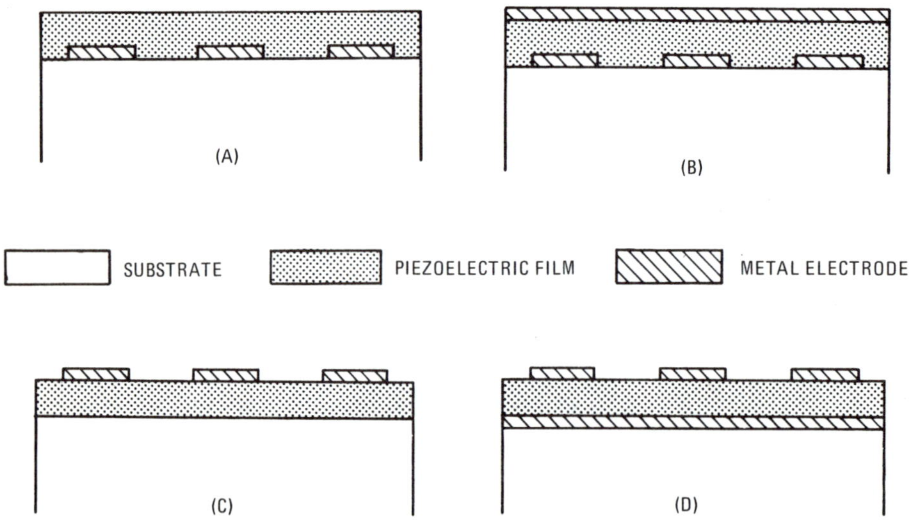

Figure 1. Piezoelectric Film Surface Elastic Wave Transducers

The theoretical coupling efficiency of the four transducer configurations for a particular film and substrate is compared in Figure 2. The quantity $\Delta v/v$[16] is plotted as a function of film thickness to acoustic wavelength ratio for a C-axis oriented ZnO film on fused quartz. It is immediately apparent that the transducer coupling efficiency is a sensitive function of both the transducer configuration and film thickness to wavelength ratio. The theoretical results indicate a point of high efficiency at low thickness to wavelength ratios (~.04) with the configurations having conducting planes at the film-substrate interface and upper film boundary. Maximum coupling efficiency occurs when the interdigital electrode is at the film-substrate interface for thickness to wavelength ratios near one-half. The coupling efficiency exceeds that obtained for interdigital electrodes on crystalline zinc oxide. The coupling efficiency for various piezoelectric film and substrate combinations give the same general relationships as those shown in Figure 2. Theoretical evidence indicates that a piezoelectric film layer has the capability of being an extremely efficient surface wave transducer structure.

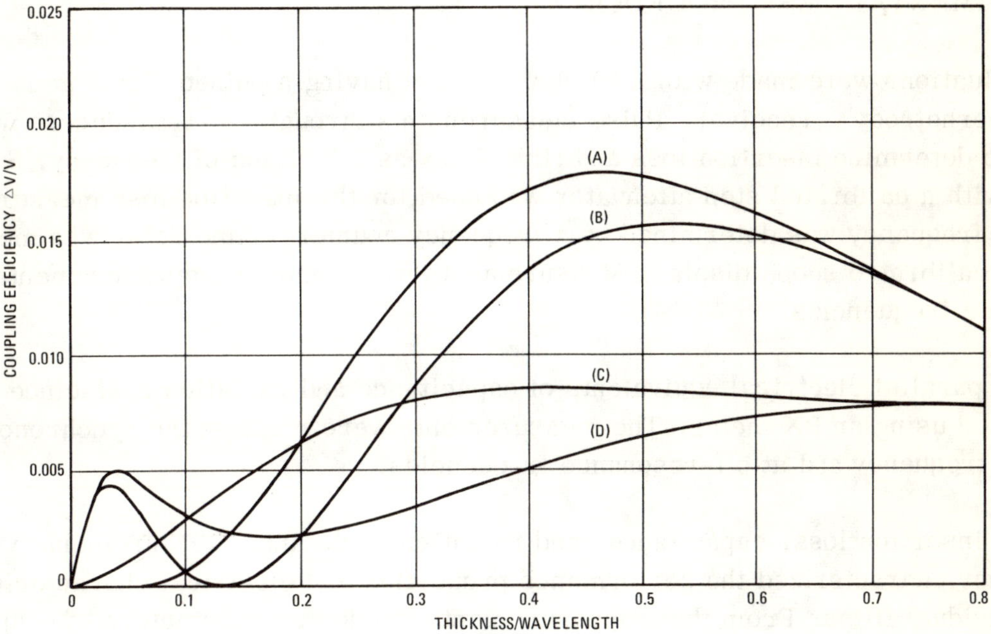

Figure 2. Coupling Efficiency of a Zinc Oxide Film Layer on Fused Quartz

TRANSDUCER FABRICATION

Three of the four configurations shown in Figure 1 were used in the experimental work. For insulating substrates and semiconductor substrates with insulating layers, (A) and (B) were used. For semiconducting substrates, configuration (C) was used. Considerable experimental data was derived using a 200 micron wavelength, 20 finger pair interdigital (ID) pattern. Individual finger width and spacing was 50 microns with a uniform finger overlap of 22 wavelengths. Two such patterns were spaced 0.65 inch apart to act as generating and detecting structures.

Polished fused quartz substrates 1" by 1" by 60 mils thick were primarily used in the experimental work. Substrates of sapphire, garnet, flint glass, silicon and silicon dioxide layers on silicon were also used. Evaporated aluminum approximately 5000 Å thick was used as the electrode metal. In a few cases, gold was used to determine electrode loading effects. The interdigital patterns were produced by standard semiconductor photoetch techniques.

The zinc oxide film layers were dc sputtered using a triode system. The films were sputtered in a 92% argon, 8% oxygen atmosphere at a pressure of 2-3 microns. The substrates were held at 175°C during sputtering and the film deposition rate was approximately 1 micron/hour. Films in the range 1-12 microns were deposited. The films were highly insulating and very adhesive to the substrate surfaces. The films were polycrystalline with the C-axis of individual grains nominally oriented normal to the substrate surface. The films were sputtered over the entire substrate surface covering both electrode and intervening region.

EXPERIMENTAL TEST PROCEDURES

Evaluations were made with a 50 ohm test-set having a pulsed signal generator and superhetrodyne receiver. Pulse measurements through two transducers were made to determine insertion loss and time delay as a function of frequency. Substitution with a calibrated step attenuator was used for the insertion loss measurement and the frequency was determined by a frequency counter. Time delay was measured from a calibrated scope display. Measurements were made at both fundamental and harmonic frequencies.

The parallel electrical equivalents of capacitance and radiation resistance were measured using an RX meter. The measurements were made at the synchronous resonance frequency and at off-resonance frequencies.

The insertion loss, capacitance, and radiation resistance measurements versus frequency characterized the performance properties and electrical characteristics for individual films. From these measurements the design parameters of coupling factor $k = (2\Delta v/v)^{1/2}$, capacitance per unit finger length and phase velocity could be calculated. The overtone measurements were useful in defining the dependence of coupling efficiency and phase velocity on film thickness.

EXPERIMENTAL RESULTS

Figure 3 gives three frequency characteristics of a 7.2 μm ZnO film on fused quartz with the 20 finger pair, 200 μm wavelength ID aluminum patterns. The loss values are total through two structures under different configuration and tuning conditions.

With only the ZnO film layer over the ID electrode (configuration A), the minimum untuned loss with 50 ohm terminations is 85 dB at 16.1 MHz. The 3 dB bandwidth is 0.6 MHz. Adding a continuous metal electrode on the ZnO layer over the finger overlap region, (configuration B), the untuned loss drops to 34 dB and the band shape is preserved. The major portion of the loss reduction is due to the substantial coupling factor improvement by altering the transducer configuration. The film thickness to wavelength ratio is 0.036 which is at the peak efficiency value for configuration B as indicated in Figure 2.

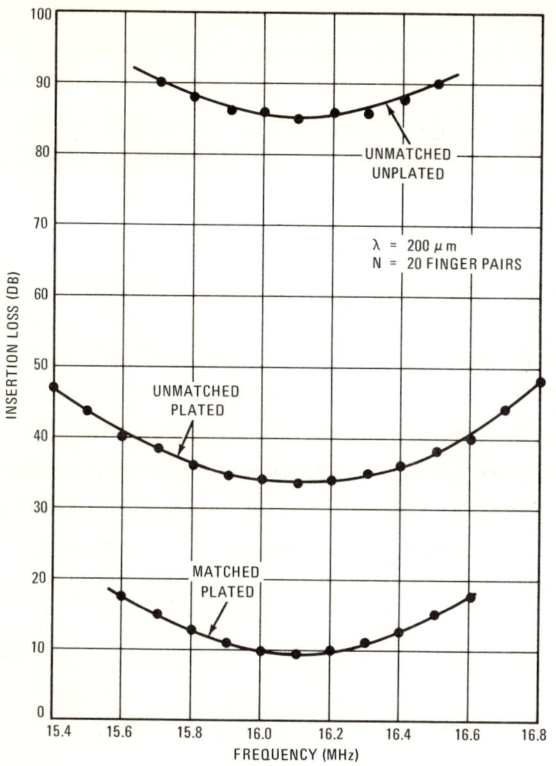

Figure 3. ZnO Film on Fused Quartz

By electrically matching the structure using a series and parallel coil the total loss could be further reduced to 9.5 dB with the bandshaped preserved. Considering 6 dB loss due to bidirectionality, single transducer efficiency is less than 2 dB for this structure.

The parallel capacitance and resistance measurements on a 3.5 μm ZnO film on fused quartz with top metal electrode are given in Figure 4. Considering that the film thickness to wavelength ratio is only 0.017, the film displays a strong electrical response near synchronism characteristic of the interdigital patterns on piezoelectric substrates. Using the Stanford equivalent circuit model[15] values of coupling factor can be calculated from the observed capacitance and parallel resistance (corrected for background film resistance).

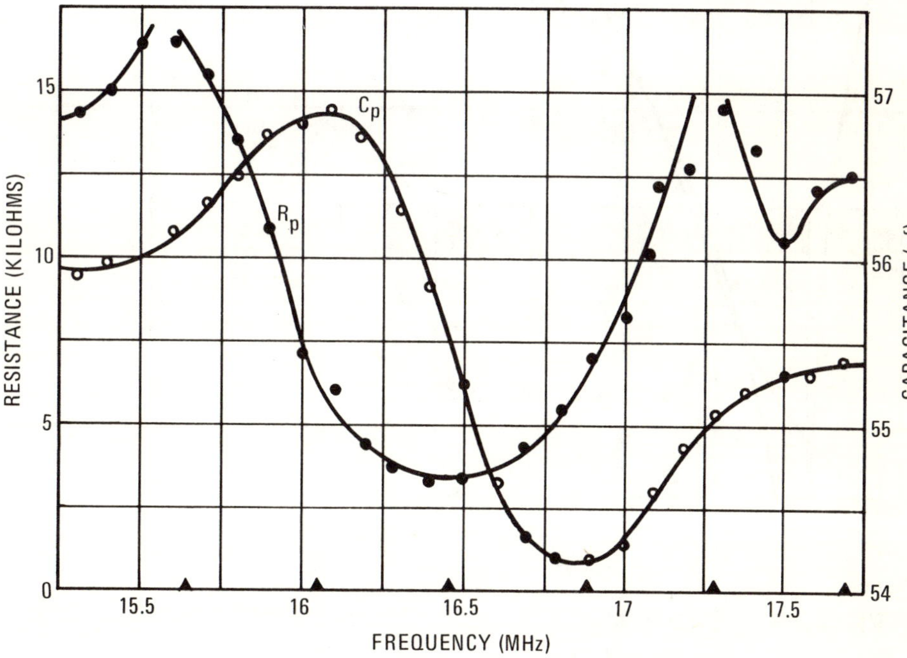

Figure 4. Parallel Capacitance and Resistance of 3.5 Micron Zinc Oxide Overlay on Fused Quartz

A coupling factor value of k = 0.039 is calculated. The same value is calculated using the observed untuned loss measurements and capacitance for the film.

Harmonic frequency-loss measurements can be used to determine the characteristic change in coupling efficiency and phase velocity with film thickness. Coupling efficiency and minimum untuned loss values are directly related in transducer circuit models. By applying a harmonic amplitude correction to the measured loss values at the harmonics a qualitative dependence of coupling efficiency on thickness to wavelength ratio is delineated. Figure 5 gives loss values for the 7.2 μm film as a function of thickness to wavelength ratio. The single conversion loss values were obtained by subtracting from the measured loss, the bidirectional loss (6 dB), the harmonic correction factor in dB (20 logN, where N is the harmonic number) and dividing the resultant loss by 2. No correction was applied for attenuation loss which becomes important at high overtone frequencies. These values are plotted for unplated and plated film layers. The dependencies follow those anticipated from theory as given in Figure 2. The region of maximum coupling efficiency is well defined. The losses are lower when the film is without a top metal electrode. This harmonic technique is extremely useful for empirically determining where maximum coupling efficiency occurs for various substrate and film combinations.

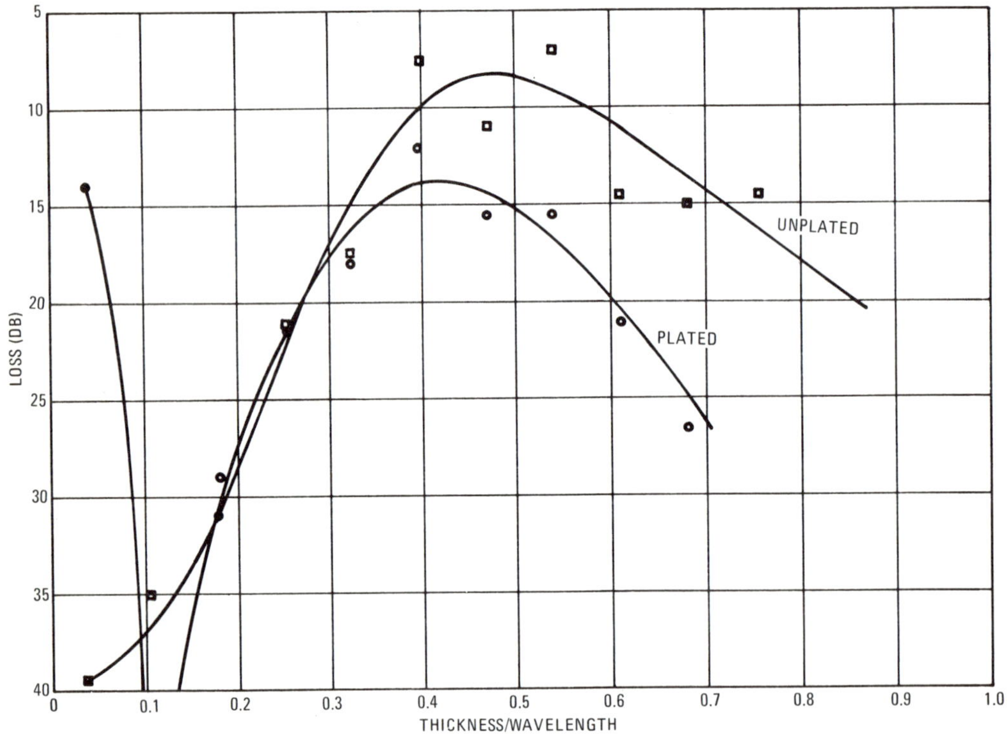

Figure 5. Conversion Efficiency Characteristic for ZnO Film on Fused Quartz

Figure 6 shows the phase velocity of zinc oxide on fused quartz for several different thickness films at fundamental and harmonic frequencies. Harmonic frequencies as high as the 29th (a frequency of 368 MHz) were generated to obtain the characteristic. The phase velocity was computed from the relation, frequency times wavelength. The frequency value was measured at the minimum transmission loss point. The wavelength was set by the transducer periodicity divided by the appropriate harmonic number. The phase velocity moves from a value of 3.4 x 10^5 cm/sec characteristic of fused quartz to a terminal velocity of 2.54 x 10^5 cm/sec. The terminal velocity value is lower than that expected from a zinc oxide single crystal substrate. The experimental curve had the same characteristic when an aluminum layer was used as a top electrode. The phase velocity is a characteristic of the elastic and piezoelectric constants of the ZnO film layers. This type of empirical curve is essential in the frequency design of film transducer structures.

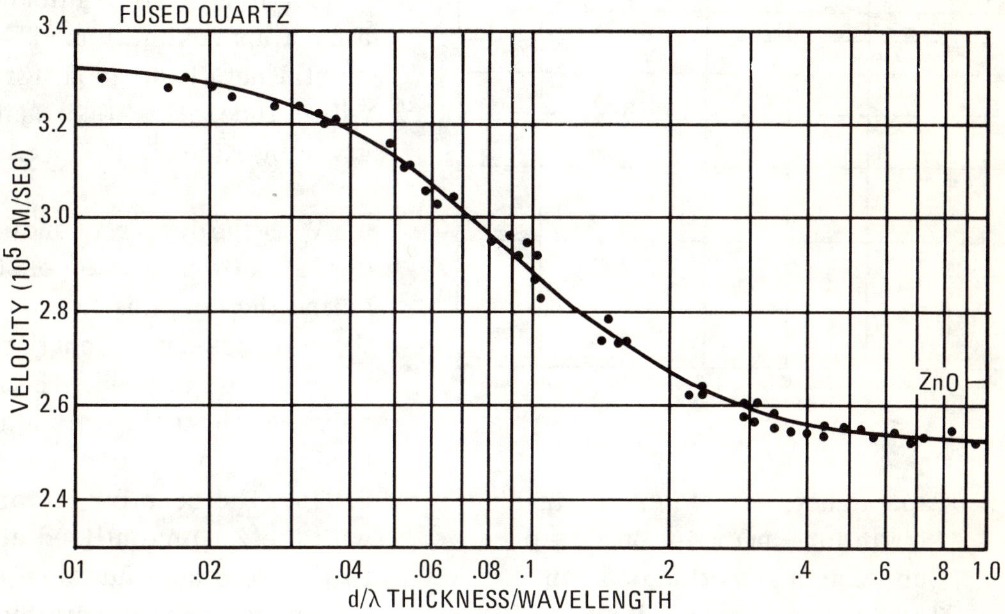

Figure 6. Phase Velocity - Zinc Oxide on Fused Quartz

In an interdigital surface wave transducer structure the capacitance and radiation resistance or coupling factor are important design parameters. These parameters are a function of the acoustic and electrical constants of the substrate, piezoelectric film and metal electrode. For the case of aluminum electrodes (light mass and thin), the design parameters need only be considered in terms of the properties of substrate and piezoelectric film. Quantitative data on capacitance and coupling factor are considered next for the ZnO/Fused Quartz structures.

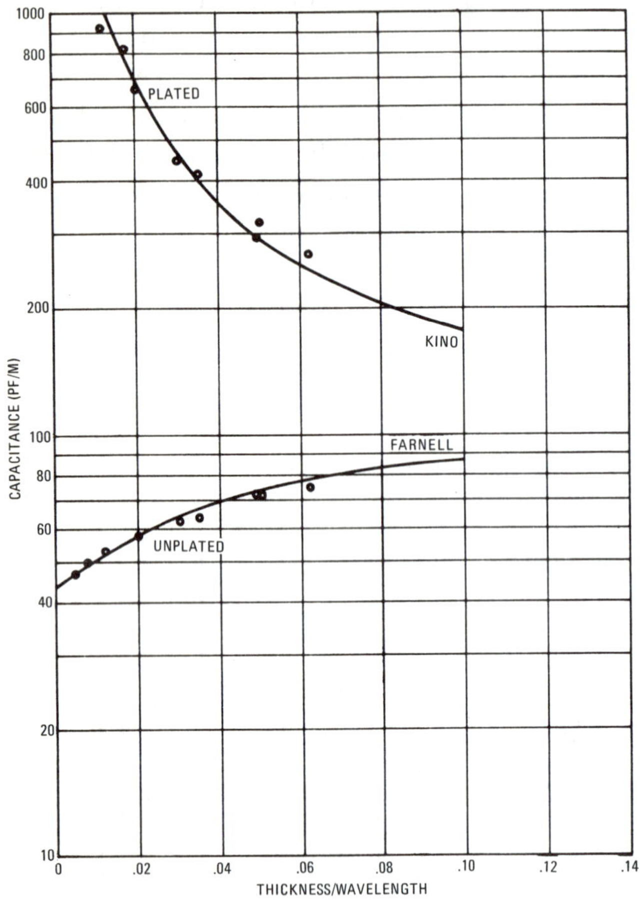

Figure 7. Capacitance Per Meter of Finger Length for ZnO Film on Fused Quartz as a Function of Film Thickness to Surface Wavelength Ratio

Figure 7 shows the capacitance relations for zinc oxide films on fused quartz as a function of film thickness. The lower curve gives capacitance per unit finger length for a film layer without the top metal electrode. The upper curve gives the values for the film layer metallized with the top electrode. The lower experimental points fit very well with the predicted theoretical expression given by Farnell.[17] The plated layer (top metal electrode) capacitance relation is approximately inversely proportional to film thickness to a first approximation. The experimental data is in good agreement with a theoretical expression given by Kino[14].

With the low frequency interdigital pattern it was possible to define the coupling factor, k, in the thickness/wavelength region from a 0.01 to 0.05. The k-values without the top metal electrode was low, on the order of 0.010. The coupling factor values with the ZnO metallized are given in Figure 8 as a function of film thickness together with theoretical. The theoretical values were taken from Figure 2 using the relation $k = (2\Delta v/v)^{1/2}$. The experimental values have the same characteristic shape as theoretical and reach a maximum value near 0.06 at a thickness to wavelength ratio of 0.035. The film transducer coupling factor is approximately 60% that of theoretical representing the single crystal condition. That theoretical values are not obtained result from the fact that the semi-random orientation of individual crystallites do not support the vibrational modes characteristic of the single crystal structure assumed in the theoretical derivations.

The basic ID pattern with ZnO films was used to determine performance, coupling factor dependencies, and transducer design parameters on flint glass, garnet, sapphire, silicon and oxided layers of silicon. The coupling efficiencies for ZnO layers on flint glass and oxided silicon substrates using configurations (A) and (B) were equal to

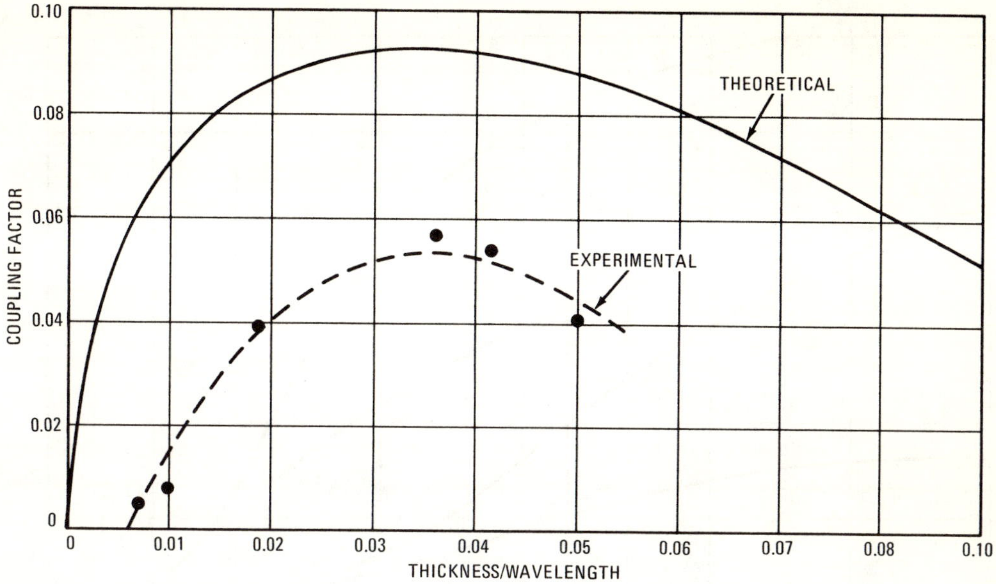

Figure 8. Electromechanical Coupling Factor of ZnO Overlays on Fused Quartz

those on fused quartz. The coupling on garnet and sapphire was below that of fused quartz although only a limited experimental sampling was made. Configuration C was used to excite surface waves on silicon. Electrically this should look like configuration D. The coupling efficiency was less than that of ZnO on fused quartz.

Phase velocity characteristics, on five of the substrate materials are shown in Figure 9. The importance of these relations in order to permit accurate frequency design of transducer structures is again emphasized. A low or high frequency dispersion characteristic can be obtained by choice of substrate and film thickness. By choosing a substrate such as flint glass, whose surface wave velocity is close to that of the film, a nondispersive transducer is obtained over a large film thickness range. The phase velocity for the silicon dioxide on silicon substrates, for which only one representative substrate has been given in the Figure, could be varied in the dispersive region by changing the oxide thickness.

CONCLUSIONS

Basic electrical and acoustic properties of piezoelectric films with interdigital electrode arrays used to generate and detect surface elastic waves has been described. The observed performance characteristics were similar to those of the common interdigital electrode on a piezoelectric substrate. The dependence of coupling efficiency capacitance and phase velocity on piezoelectric film thickness for ZnO film layers on fused quartz has been experimentally characterized. Strong coupling efficiencies can be obtained by proper selection of transducer configuration and film thickness. Methods

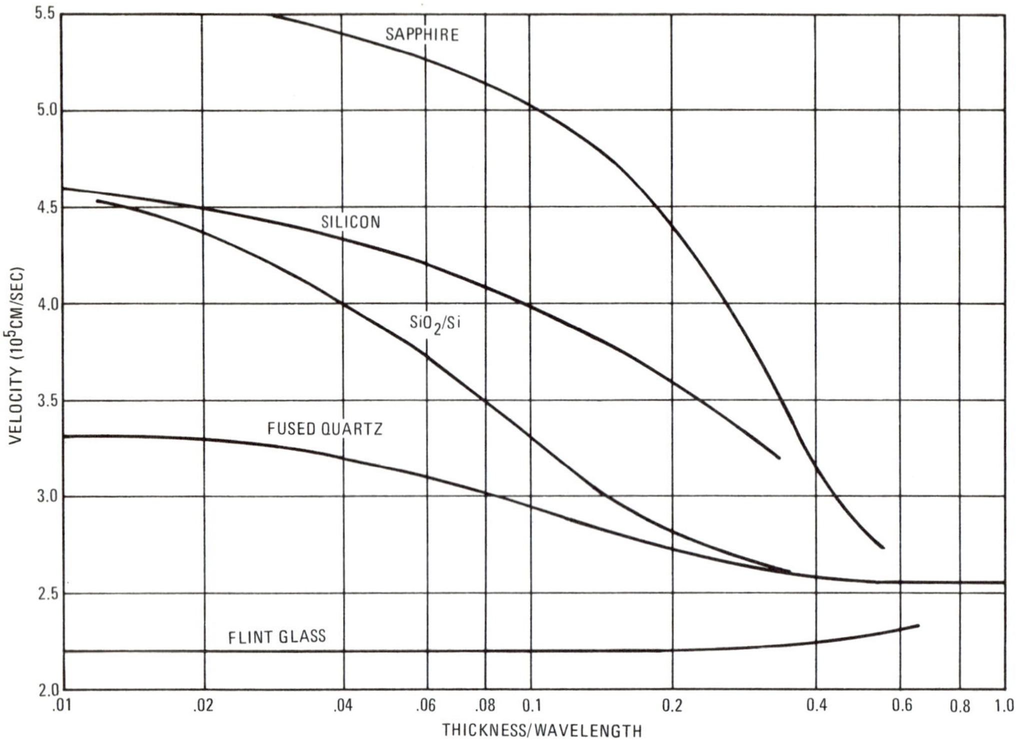

Figure 9. Phase Velocity of ZnO Film on Selected Substrates as a Function of Film Thickness to Wavelength Ratio

were described which permit the quantitative delineation of the parameters required for transducer design. These techniques can easily be extended to other piezoelectric film and substrate combinations.

The use of piezoelectric film overlays permits surface wave excitation on a wide variety of substrates while retaining many of the advantages common to the interdigital electrode on a piezoelectric. The techniques described should benefit signal processing device development particularly where it is desirable to interface with semiconductor or amorphous substrates.

ACKNOWLEDGEMENTS

The author wishes to express his appreciation to D. E. Allen, M. Adamo, J. Brewer and G. Shulda for invaluable technical assistance. The work was done under a Motorola sponsored independent development program.

REFERENCES

1. D.E. Allen, F.S. Hickernell and K.C. Wolters, "Generation and propagation of surface acoustic waves" Proceedings IEEE Region VI Conference, Phoenix, Arizona (April 1969).

2. F.S. Hickernell, "Thin film piezoelectric overlay surface wave transducers," presented at the International Surface Acoustic Wave Symposium (IBM), Yorktown Heights, New York (March, 1970).

3. F.S. Hickernell, G.F. Shulda and J.W. Brewer, "ZnO and CdS overlay surface wave transducers," presented at the 1970 IEEE Ultrasonics Symposium, San Francisco, California (October 1970).

4. R.M. White and F.W. Voltmer, "Direct piezoelectric coupling to surface elastic waves," Appl. Phys. Letters 7, 314 (1965).

5. P. Schnitzler, "Elastic surface waves in a layered system of cadmium sulfide on germanium," Appl. Phys. Letters, 11, 273 (1967).

6. H. Matthews and H. van de Vaart, "Observation of love wave propagation of UHF frequencies," Appl. Phys. Letters 14, 171 (1969).

7. R.V. Schmidt and F.W. Voltmer, "Piezoelectric elastic surface waves in anisotropic layered media," IEEE Trans MTT 17, 920 (1969).

8. M.R. Daniel and J. deKlerk, "Velocity measurements of elastic surface waves in the layered system ZnS on Al_2O_3," Appl. Phys. Letters, 16, 331 (1970).

9. P. Schnitzler, L. Bergstein, and L. Strauss, "Elastic surface waves: thin film transducers and layered-system dispersion," IEEE Trans SU. 17, 185 (1970).

10. A.P. van den Heuvel, "Surface wave electronics." Science and Technology, No. 85, 52 (January 1969).

11. R.M. White, "Surface elastic waves" Proc. IEEE, 58, 1238 (1970) (see page 1248).

12. L.P. Solie, "Piezoelectric acoustic surface waves for a film on substrate" Appl. Phys. Letters 18, 111 (1971).

13. W.R. Smith, "Excitation of surface acoustic waves with piezoelectric film overlays," 1970 IEEE Ultrasonics Symposium, San Francisco, California (October 1970).

14. G.S. Kino, M. Heckman, L. Solie, and D. Winslow, "A theory for interdigital rayleigh wave transducers on a nonpiezoelectric substrate" 1970 IEEE Ultrasonics Symposium, San Francisco, California (October 1970).

15. W.R. Smith, H.M. Gerard, J.H. Collins, T.M. Reeder and H.J. Shaw, "Analysis of interdigital surface wave transducers by use of an equivalent circuit model," IEEE Trans. MTT 17, 856, (1969).

16. J.J. Campbell and W.R. Jones, "A method for estimating optimal crystal cuts and propagation directions for excitation of piezoelectric surface waves," IEEE Transactions on Sonics and Ultrasonics 15, 209 (1968).

17. G.W. Farnell, I.A. Cermak, P. Silvester, and S.K. Wong, "Capacitance and field distribution for interdigital surface wave transducers" IEEE Trans. Sonics and Ultrasonics, 17, 188 (1970).

WAVEGUIDES AND DISTRIBUTED COMPONENTS FOR ELASTIC SURFACE WAVES

Robert M. Knox and Donald B. Owen
IIT Research Institute
Chicago, Illinois 60616

INTRODUCTION

Investigations into the use of bulk elastic waves for delay lines have brought about the realization of useful delay line devices. The need, however, for greater versatility in delay lines has encouraged research in the use of surface waves for this application. Publications by White[1] and Tiersten,[2,3] and others[4,5,6,7] have shown that the elastic surface wave can be guided by narrow superficial films. The "guidability" of such films is determined by the mechanical properties of the film and substrate as well as the transverse dimensions of the film.

Tiersten has provided a theoretical analysis for two types of guides, a strip guide in which a narrow film is deposited along the desired path in the direction of propagation and a slot guide in which the film is deposited everywhere except along the desired propagation path. Although the slot guide shows promise of being less dispersive, most component investigations have thus far employed the strip guide. Figure 1 shows the strip and slot guide configurations and Fig. 2 the dispersion characteristics derived by Tiersten. Those results have been verified for both the strip[4] and slot[5] guide.

Adkins and Hughes,[6,7] using an approach similar to Tiersten, have also derived the dispersion characteristics for a thin film guide. Experimental verification of this analysis for the specific case of gold films on fused quartz was obtained.

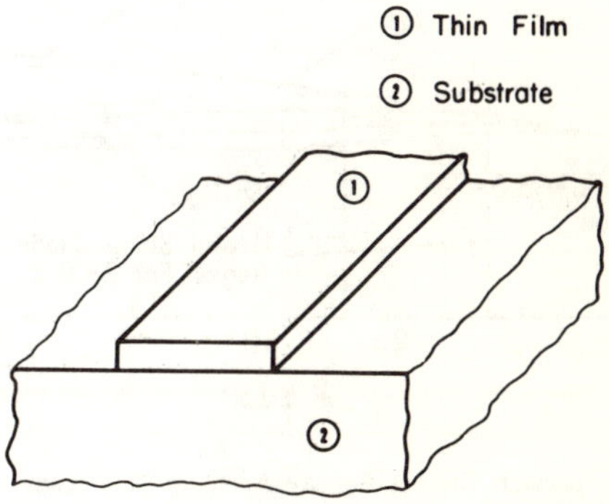

1a) Acoustic Film Strip Guide

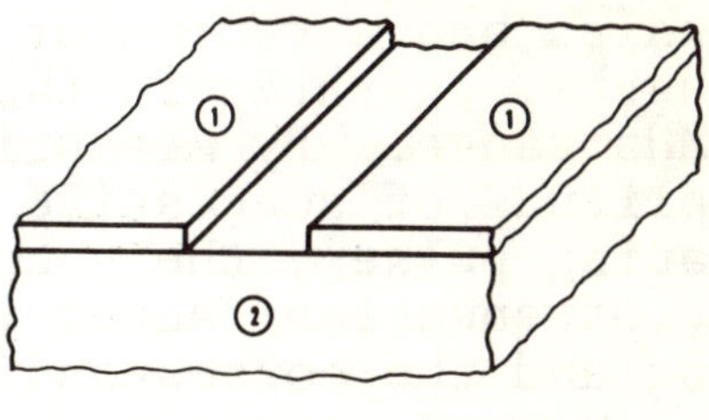

1b) Acoustic Film Channel Guide

Adkins and Hughes also present a theoretical model for the case of coupled guides[7] for which experimental correlation with theoretical prediction of coupling length was obtained.

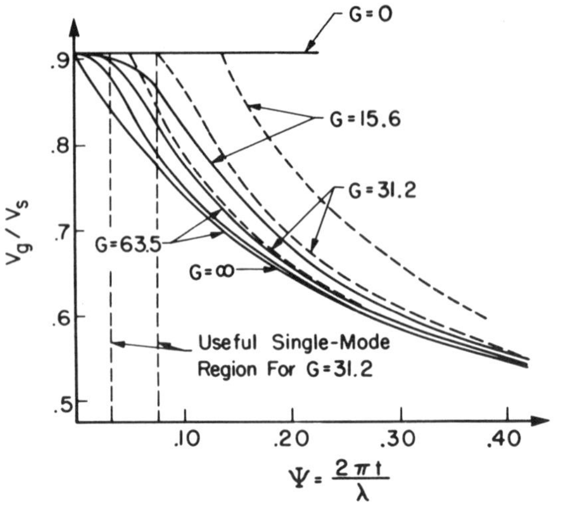

2a) Dispersion Curves For An Acoustic Guide Formed By A Gold Film Strip On Fused Silica

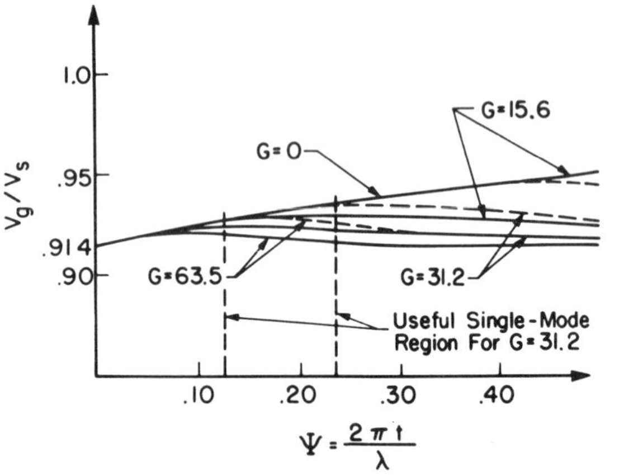

2b) Dispersion Curves For An Acoustic Slot Guide Formed By An Aluminum Film On T-40 Glass

Knox and Owen[5] have fabricated a ring resonator with a gold strip on fused quartz. The device, which had a radius of 15 wavelengths, operated in the 20-30 MHz range. By measuring the amplitude of successive circulating pulses, the coupling factor, attenuation factor in the ring and the corresponding unloaded Q were determined.

Described herein is additional experimental work with ring resonator circuits also designed for operation around 30 MHz. This frequency range was selected for ease of fabrication, but the results are largely translatable to higher frequencies except for the consideration of losses which increase with frequency.

The circuits were fabricated from thin films of gold on a fused quartz substrate. The method of surface wave generation was the Hybrid[9] technique in which an interdigital comb-like array is fabricated on the substrate (which is nonpiezoelectric) and a single crystal block of a piezoelectric material is placed over the array. A high frequency signal applied to the array launches a surface wave in the block which is simultaneously coupled into the substrate. The coupling is achieved by using a very thin liquid layer between the block and the substrate.

A method is described whereby CW measurements were made using CW pulses. The use of pulses circumvents the difficulty of electromagnetic pickup directly from input to output. The Q of the resonator was determined from the CW frequency response.

EXPERIMENTAL RESULTS

Three ring resonator circuits of the configurations shown in Fig. 3 were fabricated.

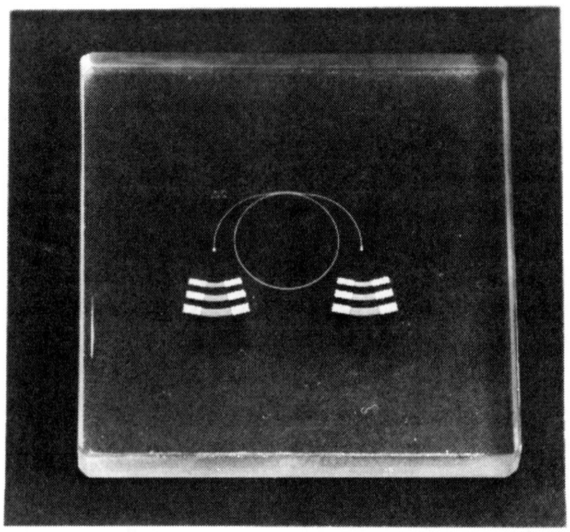

Fig. 3 View of a Guided Surface Wave Ring Resonator. Radius of the ring is $40\lambda g$.

The radii of these circular waveguide resonators were 15λg, 25λg and 40λg where λg is the wavelength of the elastic wave in the guide at 30 MHz. The circuit shown is the 40λg radius resonator with a new type of interdigital hybrid transducer array. The fingers of the transducer were circular so that the acoustic waves were focussed into the end of the waveguide as described by Engan.[10] This design has proven to be considerably more efficient than earlier transducers which had straight fingers and used a horn-type accumulator to guide the energy into the end of the waveguide.[5]

The coupling region of the circuit was designed using the results provided by Adkins and Hughes.[7] It was designed to be one coupling length at 30 MHz, which is 17.1 guide wavelengths or 1.62 mm long. The guide width and thickness were .051 mm and 1.78μm respectively.

Pulse measurement techniques have been employed up to this time because of the difficulty of making CW measurements in the presence of electromagnetic coupling from input to output. The output (from a square-law detector) of the 40λg ring resonator when excited by a single narrow R.F. pulse is shown in Figs. 4(a) and 4(b). Figure 4(b) shows an expanded portion of the output from 140μs to 230μs. It can be seen that without benefit of acoustic amplification useful signal output can be obtained to approximately 200μs.

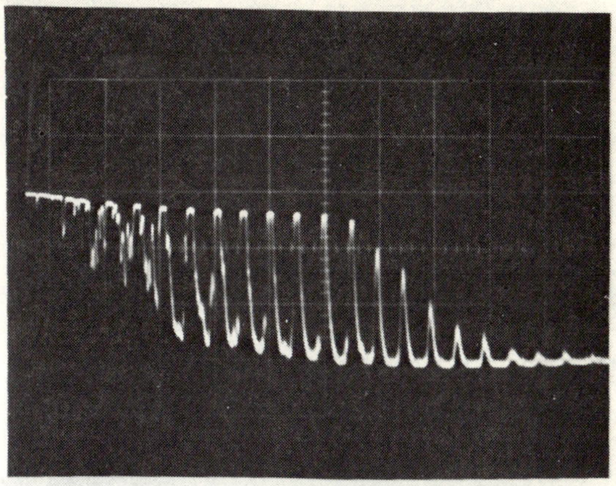

Fig. 4(a) Amplified Output of the Ring Resonator for a Single R.F. Input Pulse. (Amplifier Saturated for the First 120μs. Horizontal Scale: 20μs/div.)

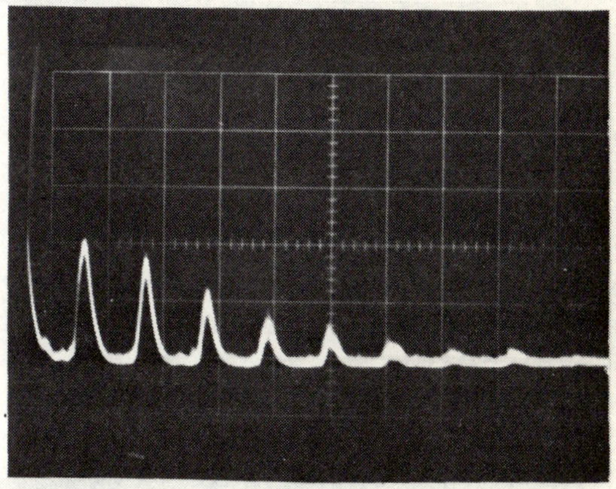

Fig. 4(b) The Amplified Ring Resonator Output Shown from 140 to 230μs. Horizontal Scale: 10μs/div.

Examination of the first three acoustic pulses arriving at the output yields information about the power coupling factor and the wave attenuation in the ring. A typical output is shown in Fig. 5.

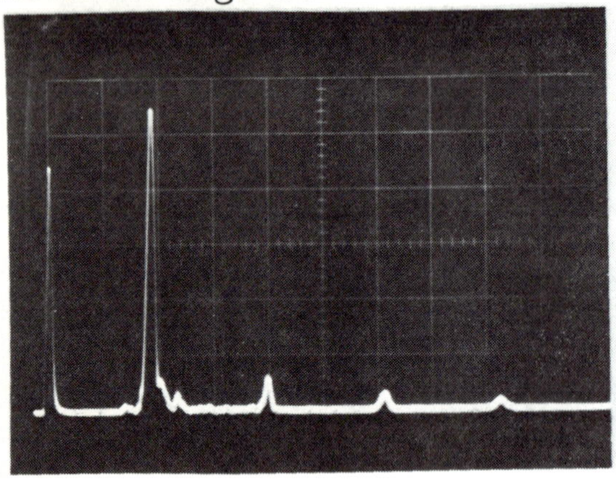

Fig. 5 Initial Output Pulses from the Ring Resonator. Horizontal Scale: 5μs/div.

The first pulse on the left is electromagnetic pickup. The second pulse from the left is the acoustic wave which travels directly through the guide without circulating. The third pulse (at 20μs delay) is that portion of the pulse which has made one circulation of the ring before being coupled back out into the guide.

The ratio of the first circulating pulse P_1 to the non-circulating pulse P_o is

$$\frac{P_1}{P_o} = \frac{K^2 e^{-\alpha C}}{1 - K} \qquad (1)$$

where K is the power coupling coefficient, α is the attenuation factor in the ring waveguide, and C is the circumference of the ring. The ratio of any two subsequent output pulses is given by:

$$\frac{P_n}{P_{n-1}} = (1 - K)\, e^{-\alpha C} \qquad (2)$$

Thus, from these two equations, it is possible to determine α and K from the two indicated ratios. The attenuation factor and the corresponding unloaded Q of the resonator are shown in Table I for each of the three resonator circuits. The Q was calculated using the expression

$$Q = \frac{\pi}{\alpha \lambda g} \qquad (3)$$

The attenuation factor in the two largest waveguide rings should be about the same. The difference may be due to film edge definition or some other aspect of guide uniformity. For smaller radii such as 15λg, an increase in α might be expected due to radiation losses. The increase here, however, does not seem to be particularly large.

A typical coupling curve for a surface wave directional coupler has been given by Adkins and Hughes[11] and is shown in Fig. 6. This curve is the ratio of the power in the coupled arm divided by the power transmitted to the output. In Fig. 7, the coupler characteristics for the coupler in the 40λg ring resonator circuit are shown. The coupling factor is the coupled power divided by the input power. Considerable similarity can be seen between the curves of Figs. 6 and 7. Since the terminals of the ring resonator circuit were not necessarily well matched, it is not unexpected to observe nonuniformity in the coupler characteristics of Fig. 7.

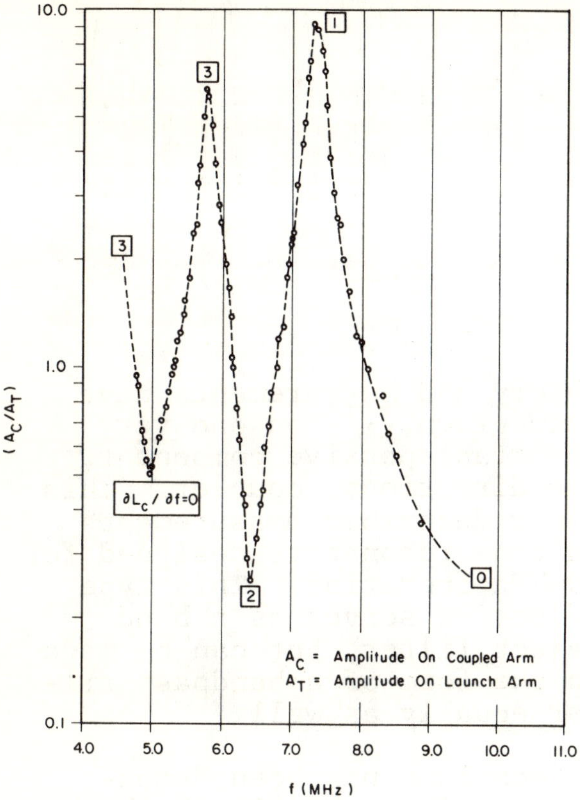

Fig. 6 COUPLER CHARACTARISTICS FOR A TYPICAL ACOUSTIC SURFACE WAVE COUPLER

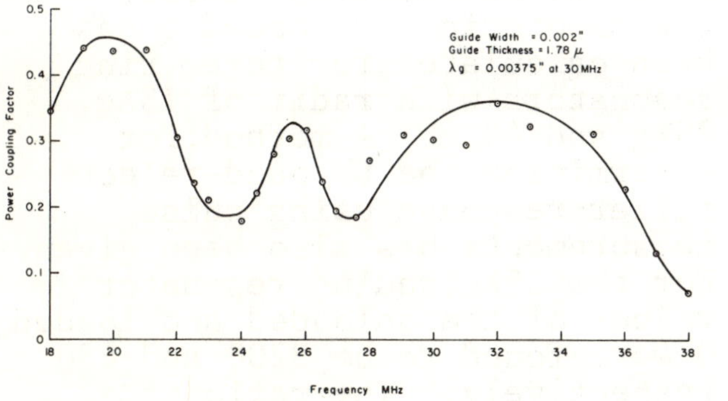

Fig. 7 COUPLING FACTOR FOR A SURFACE WAVE RING RESONATOR WITH RADIUS OF 40 λg AND DESIGN COUPLING LENGTH OF 17 λg AT 30 MHz

A method has been devised for determining the CW response of the ring resonator which, of course, forms a band reject filter. Pulse excitation is employed in order to avoid electromagnetic pickup. If the length of the pulse is extended to 1.5 times the delay time in the ring waveguide, then waves from the leading portion of the pulse which has circulated the ring can interfere with waves in the trailing portion of the pulse. As the frequency is varied the waves will constructively and destructively interfere. Filtering takes place during destructive interference.

The sequence of photos in Fig. 8 of the ring resonator output with a 15μs wide pulse show the destructive and constructive interferance as the frequency is increased from 26.980 to 27.172 MHz. Observing the output for example at the center verticle graticule, the level goes from minimum (maximum rejection) to half-power point, to maximum, to half-power point and back to minimum. The rejection bandwidth and loaded Q can be determined from the frequency interval between half power frequencies on each side of a minimum. The periodic band-rejection response of the 25λg ring resonator near 27 MHz is shown in Fig. 9. The loaded Q of this filter was determined to be about 480 from these measurements.

CONCLUSIONS

The objective of current investigations into acoustic surface waveguides is to ultimately achieve a technology which will permit RF signal processing using distributed parameter elements in a manner similar to current microwave practice.

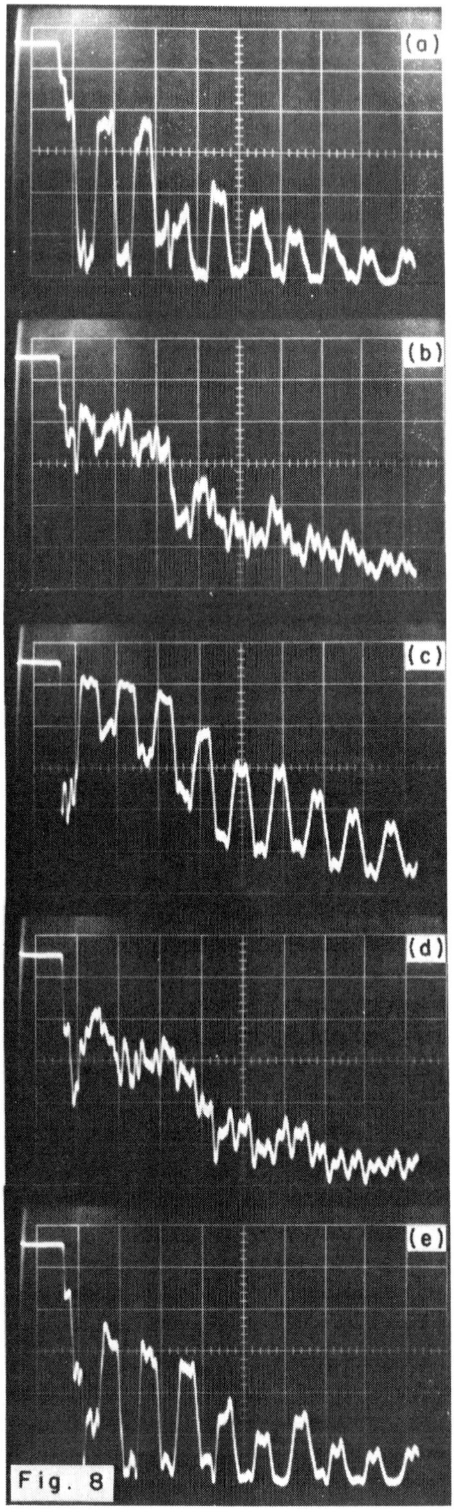

Fig. 8 Pulse Interference for Successively Increasing Frequencies.

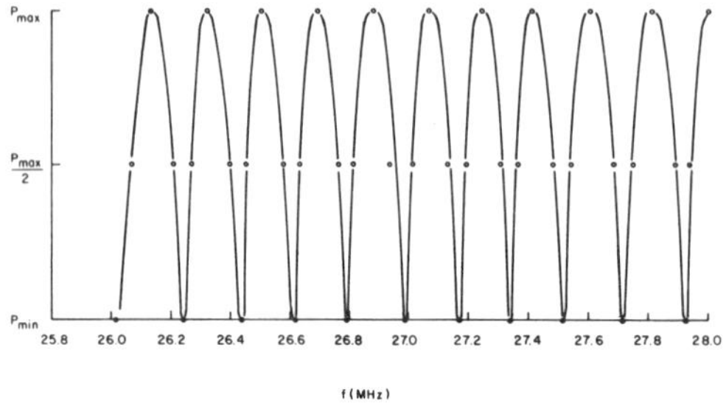

Fig. 9 PARTIAL FREQUENCY RESPONSE OF A 25λg RADIUS RING RESONATOR

Theory and measurements have been given,[3,11] on one very important passive component, the directional coupler. This paper described measurements on ring resonators designed for 30 MHz operation. This type of device serves as a band reject filter, but can be made in the form of a bandpass filter equally as well.

A technique has been demonstrated for measuring both the attenuation factor and coupling factor of a ring resonator. From the attenuation factor, the theoretical unloaded Q has been calculated for three ring resonators with radii of 15λg, 25λg and 40λg. A method for determining the CW band-reject filter response using pulse measurements has also been given. For the 25λg radius resonator values of the unloaded and loaded Q were found to be 3200 and 480 respectively. The ratio of unloaded to loaded Q of 6.6:1 is reasonable for the coupling coefficient (0.2 - 0.3) of these resonators.

The CW response of the resonator shows that resonant responses are very close together (approximately 175 kHz separation) due to the very large (in wavelengths) radius of the ring. This separation is increased as the ring radius is reduced. However, the losses increase as the radius decreases so that the Q of the resonator also decreases.

Therefore, a compromise between useful spurious-free bandwidth and high Q is required. The spurious responses, however, may be removed by using very narrow band transducers.

ACKNOWLEDGEMENTS

The authors wish to acknowledge the helpful discussions with Drs. A.P. van den Heuvel, P.P. Toulios and H.F. Tiersten during the course of this work, and the assistance of Mr. R. Richter in the fabrication of the circuits.

TABLE I. ATTENUATION FACTOR FOR THE RING RESONATOR CIRCUITS

Ring Radius		Attentuation Factor α		
λg	(cm)	db/CM	db/λg	Q
15	(.143)	2.5	0.0238	1140
25	(.238)	0.9	0.0085	3200
40	(.381)	1.6	0.0152	1790

REFERENCES

1. D.L. White, IEEE Symposium on Sonics and Ultrasonics, Vancouver, Canada, October 1967.

2. H.F. Tiersten, IEEE Symposium on Sonics and Ultrasonics, Vancouver, Canada,1967.

3. H.F. Tiersten, J. of Applied Physics, 46, 8, pp. 770-789, February, 1969.

4. R.M. Knox and A.P. van den Heuvel, European Microwave Conference, London, England, September 8-12, 1969.

5. R.M. Knox and D.B. Owen, "Distributed Components For Microwave Elastic Surface Wave Circuits", Digest of the Symposium in Microwave Theory and Techniques, Newport Beach, California, May 11-14, 1970, pp. 370-374.

6. L.R. Adkins and A.J. Hughes, IEEE Trans. on Microwave Theory and Techniques, MTT-17, pp. 904-911, November, 1969.

7. L.R. Adkins and A.J. Hughes, "Acoustic Surface Wave Directional Couplers", Digest of the Symposium on Microwave Theory and Techniques, Newport Beach, California, May 11-14, 1970, pp. 375-379.

8. H.F. Tiersten, National Electronics Conference, Chicago Illinois, December 7-9, 1970.

9. A.P. van den Heuvel, et al, Paper presented at the 1969 IEEE Ultrasonics Symposium, St. Louis, Mo., September, 1969.

10. H. Engan, Experiments with Elastic Surface Waves in Piezoelectric Ceramics, Electronics Research Lab Report TE-128, The Norwegian Institute of Technology, July, 1969.

11. L.R. Adkins, et al, "Microwave Magnetic and Acoustic Materials, Microwave Surface Wave Techniques", Research and Development Technical Report, ECOM-0321-1, March, 1970.

WAVES IN A LAYERED STRUCTURE OF SOLID-LIQUID-SOLID --- APPLICATION TO SURFACE WAVE PROPAGATION AND AMPLIFICATION

Professor Wen-Chung Wang
Electrical Engineering Department
Polytechnic Institute of Brooklyn
Brooklyn, New York

I. Introduction:

In the area of microacoustics, interest in the liquid layer structure is mainly aroused by the observation of high conversion efficiency between the Rayleigh wave and the layer wave or the buck wave.[1-9] So far the frequency of operation has been observed up to several hundred megacycles. In this paper, we first give an intuitive physical understanding of the layer structure and then proceed with the experiment, the theory and its potential applications.

Figure 1 is the basic structure. Two solid plates are separated by a liquid layer. The upper solid is with a shear wave velocity c_t and the lower with c_t'. $c_t < c_t'$. The types of waves propagating in such a structure cannot be

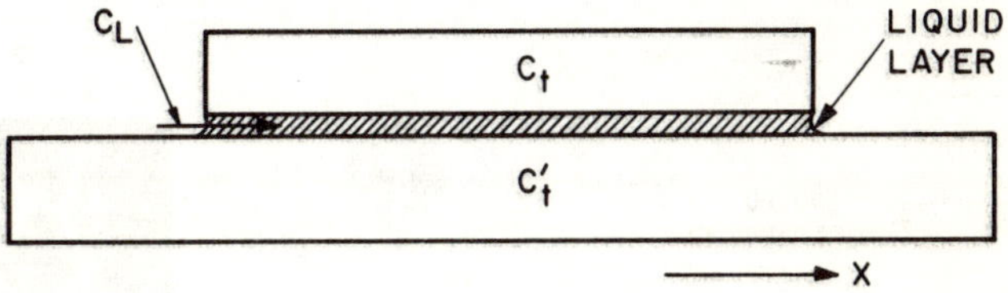

Fig. 1 - Basic structure; a liquid layer sandwiched between two solids. c_L; x-directed sonic phase velocity in the liquid. c_t; shear wave velocity of upper solid. c_t'; shear wave velocity of lower solid. $c_t < c_t'$.

known until a set of rather involved equations is analyzed. However, if the x-directed sonic velocity in the liquid layer c_L is assumed to be a known quantity, one can learn about the behavior of the wave propagation. The value of c_L in general, depends on both the thickness of the layer and the frequency. According to the phase matching condition for the case of $c_L < c_t$ and c_t', the sonic wave is bounded in the liquid layer and evanescent in the adjacent solids and the energy penetration will be deep in the solid with a slower velocity c_t (upper) but slight in the solid possessing a higher velocity. For $c_t < c_L < c_t'$, the sonic wave will not be bounded in the layer but continuously leaks energy into the upper solid as it propagates. The leaky wave will propagate in the form of transverse wave. It is then clear that in the first case c_L is a real quantity and in the second, c_L is a complex quantity. The imaginary part of c_L will determine the leaky rate of sonic wave in the layer.

II. <u>Experiments:</u>

For clarity, experiments will be presented in three parts. Part (A) will describe a case where only the leaky wave exists. Part (B) reports a case of bounded wave and part (C) describes the case where either the bounded or the leaky may prevail.

(A) Leaky Wave: Figure 2a shows a quartz surface wave delay line which will be used as the lower substrate of the layered structure. Two 30 MC interdigital transducers are deposited on the quartz substrate. The response at the output transducer is shown in Fig. 3a. The r.f. pulse appearing at 0 μsec, designated by I, is the input signal radiating through the air. Pulse II, appearing ≈ 5.6 μsec later, is the Rayleigh wave detected at the output transducer. Figure 2b shows that a drop of silicone fluid (water, acetone, and grease, etc., give similar effect) placed in the path of the Rayleigh wave. Figure 3b is the corresponding output, where the Rayleigh wave, II, is seen greatly attenuated. Figure 2c shows that a piece of CdS of dimension 0.88 x 0.6 x 0.25 (cm^3) is placed on the top of the silicone fluid forming a very thin layer. The corresponding acoustic output is shown in Fig. 3c, and a new pulse III, is seen appearing ≈ 2 μsec after the pulse II. Before we explain the new pulse III and some of its characteristics, it is necessary to understand the implication of a very thin fluid layer. A thin fluid-layer system is a

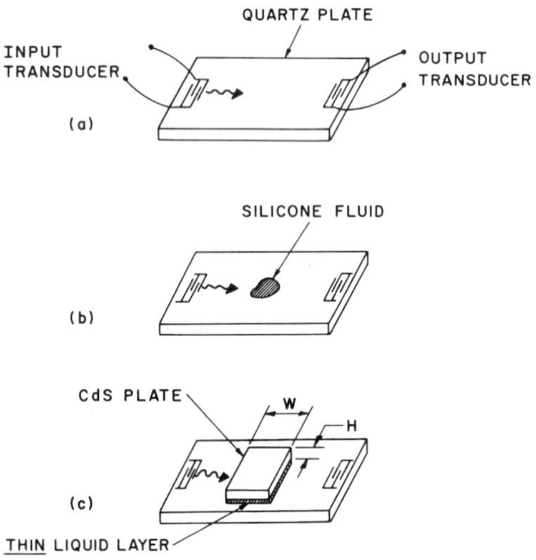

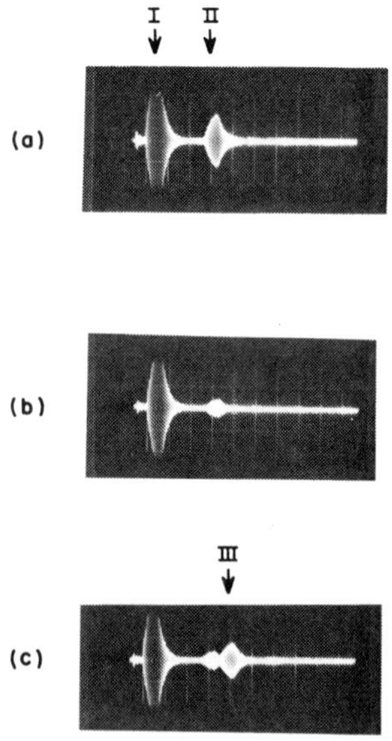

Fig. 2 - Experimental configuration. (a) X-cut quartz surface wave delay line, (b) a drop of silicone fluid is placed on the path of wave propagation, (c) A CdS plate is placed on top of the silicone fluid.

Fig. 3 - rf signal detected at output transducer corresponding to various condition of Fig. 2. (a) "I" designates the input 30 rf pulse radiating through the air, "II" rf pulse output of the Quartz delay line, (b) rf pulse II is greatly attenuated by a drop of silicone fluid (SF-81), (c) a new rf pulse III is appearing due to the presence of CdS plate.

system in which $H/\lambda_F << 1$, H is the layer thickness and λ_F is the sonic wavelength in the fluid. When H = 0 and no fluid is applied, the velocity observed experimentally is the surface wave velocity of the quartz substrate. Therefore, it seems reasonable to assume that the layer velocity c_L approaches to the surface wave velocity under the condition of $H/\lambda_F << 1$, (experiment described later will confirm its validity) which in fact says that the transverse wave will propagate in the CdS plate. The direction of its propagation with respect to the plate surface normal is $\theta = \sin^{-1}(c_t/c_L)$. Since the CdS crystal is a parallel plate, the propagating transverse wave will be reflected back from top surface of the plate and reconvert into surface wave at the liquid interface.

A modified structure of Figs. 1 and 2c is shown in Fig. 4a in which a piece of double-sided scotch tape is used to replace the middle portion of the liquid layer. Without the presence of any liquid couplant in the modified structure no output signal, as shown in Fig. 4b, is observed. Figure 4c is the corresponding output response of Fig. 4a. Since surface wave would be totally attenuated in the presence of the scotch tape, the finding on the modified structure confirms the existence of the leaky wave and offers an indication on the leaky rate.

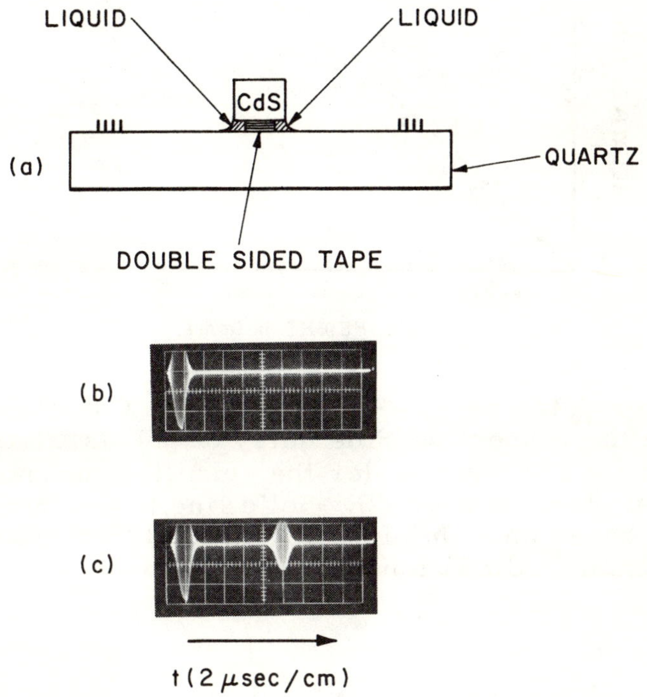

Fig. 4 - (a) Modified structure of Fig. 1. The middle portion of the fluid layer is replaced by a piece of double sided tape. 30 MC delay line is used here.

(b) No output signal is observed when both the liquid parts are removed from the modified structure.

(c) Output response of the modified structure (a).

Another confirmation on the existence of the transverse wave and its direction of propagation comes from Fig. 5 which gives the data of relative time delay $\tau = \tau_{III} - \tau_{II}$ vs. the CdS plate height h. τ_{III} and τ_{II} are the respective time delay of pulses III and II in Fig. 3. The experimental data are taken under the condition that the plate width is long enough that no multiple reflection is allowed to take place inside the CdS crystal. The experiment data (x) are seen in general agreement with the calculation (lines made based on the simple formula that the angle of propagation with respect to the surface normal is $\theta = \sin^{-1}(c_t/c_L)$.

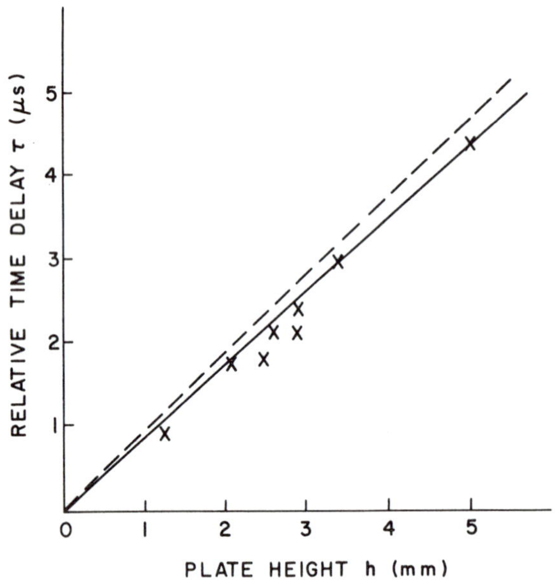

Fig. 5 - Relative time delay $\tau = \tau_{III} - \tau_{II}$ versus the CdS plate height H. τ_{III} and τ_{II} are the respective time delay of pulses III and II. Experimental data (x) are taken under the condition that no multiple reflection is taken place inside CdS. Solid line (——) represents calculations made based on assuming $c_L = 2.8 \times 10^5$ cm/sec. Dotted line (---) represents calculations made based on $c_L = c_t = 3.15 \times 10^5$ cm/sec.

Finally, we report the dispersion measurements made on several thin layer structures; (i) the CdS-liquid-quartz structure, frequency from 150 to 175MC, from 24 to 34MC and (ii) the CdS-grease-$LiNbO_3$, from 90 to 120MC. In all cases no appreciable dispersion are detected. Figure 6 presents a typical set of pictures to illustrate the measurement, for which the substrate is a 30 MC quartz surface wave delay line. The low dispersion characteristics could be attributed to the following two reasons: (i) Under the condition of thin-layer thickness (H, estimated $\simeq 0.2\mu$) the leaky rate is high that the distance which the sonic wave travels in the layer is short, (say, $\simeq 20\lambda_F$) and (ii) the change in phase velocity is small due to the fact that the value of H/λ_F remains quite small in the frequency range of measurement.

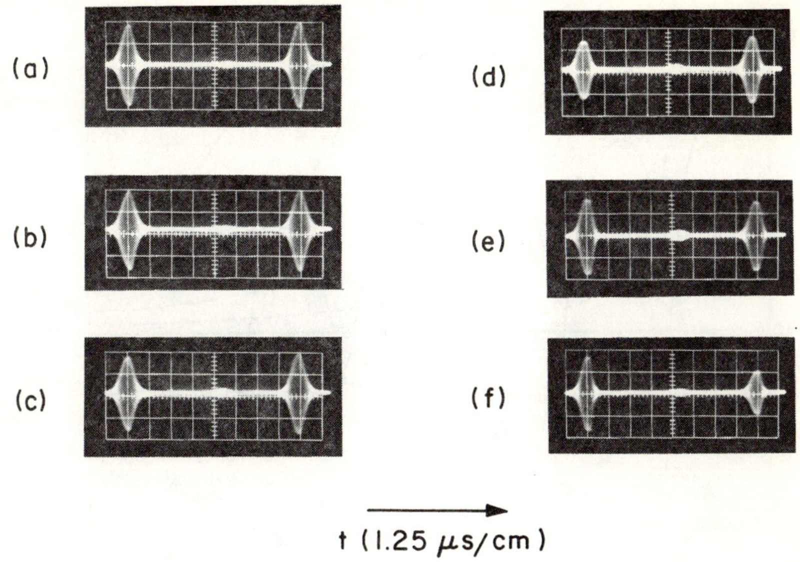

Fig. 6 - Dispersion Characteristics: The center frequency of the surface delay line is at 30 MC. The CdS overlay dimensions; $1(\ell) \times 0.5$ (h) $\times 0.5$ (w) cm^3. Pictures are taken at the following frequencies: (a) 24 MC (b) 26 MC (c) 28 MC (d) 30 MC (e) 32 MC and (f) 34 MC.

(B) Bounded Wave: When the upper solid in Fig. 2c is identical or elastically similar ($c_t' \simeq c_t$ etc.) to the lower quartz substrate, the layer velocity c_L will be less than the shear wave velocity of both the adjacent solids. In this case, the wave will propagate in the fluid and evanescent in the solids. Figure 7 shows the experimental data, represented by (·), which is seen in general agreement with our bounded wave calculation (lines). The theoretical formulation of the problem will be discussed in Section III. In the experiment soda-lima glass is used for the upper solid. The acoustic properties (ρ, c_t, c_p) of the glass in the literature vary but agree, within 10% of those for the lower quartz substrate.

(C) Leaky or Bounded Wave: Here the thickness of the liquid layer in Fig. 2c is no longer restricted to very thin. The layer velocity which is a function of the layer thickness can be varied from the velocity of sound in water to that of surface wave of quartz. The type of wave propagation in the composite structure is then determined by the layer thickness. By adjusting the layer thickness of Fig. 2c both types of the wave, the leaky and the bounded, have been observed. It is understood that so long as the thickness of the CdS plate is larger than the depth of penetration of the evanescent waves, the plate thickness does not effect the performance of bounded wave.

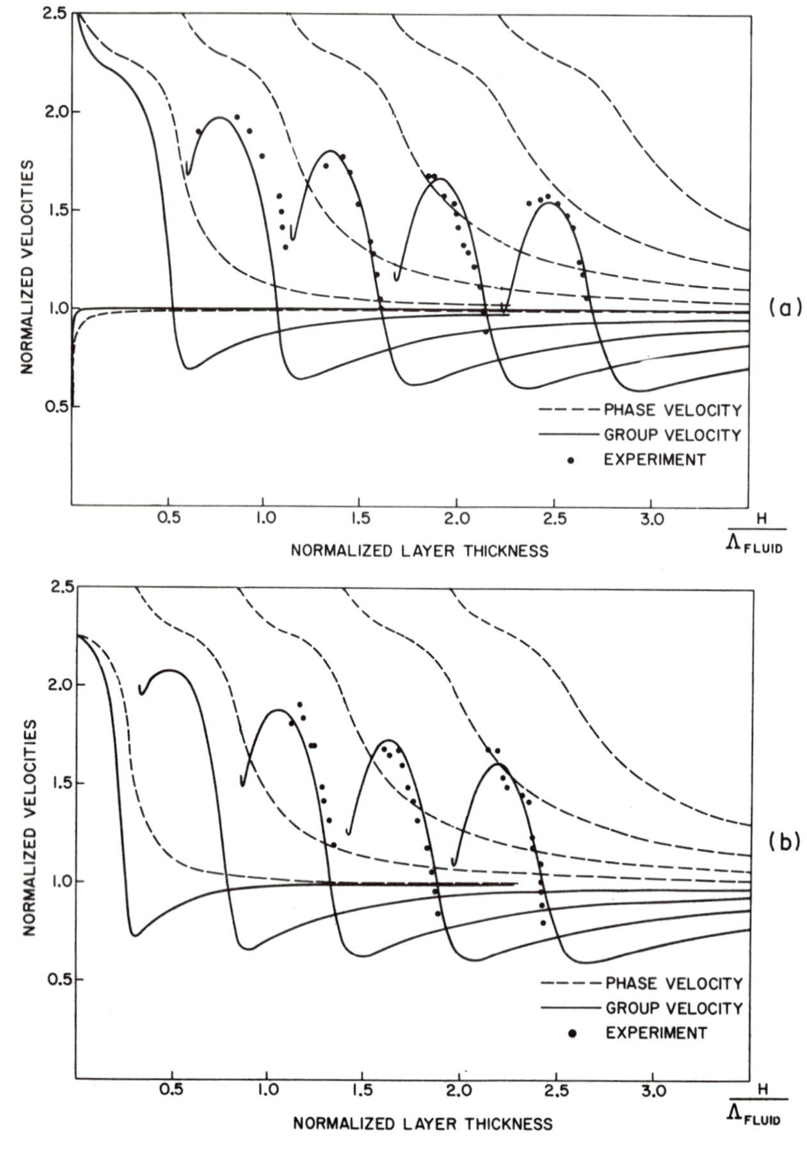

Fig. 7 - (a) Velocities of propagation normalized to fluid velocity, even symmetry;

(b) Velocities of propagation normalized to fluid velocity, odd symmetry.

III. Theory:

The theoretical model of Fig. 1 is shown in Fig. 8, two semi-infinite isotropic solids, each with differing elastic constants (λ_s, μ_s, λ'_s, μ'_s) and mass densities (ρ_s, ρ'_s) are separated by a distance 2H. The intervening space is filled with an ideal fluid of λ_F and ρ_F. The coordinate origin is chosen to lie in the fluid layer midway between the two solids with an x-axis oriented horizontally in the direction of propagation, and the z-axis vertically upwards. We seek solutions for waves which propagatte in the x-direction, obtaining the z-dependence of wave motion from the elastic wave equations which must be satisfied in the fluid layer and adjacent solids. No y variation is assumed.

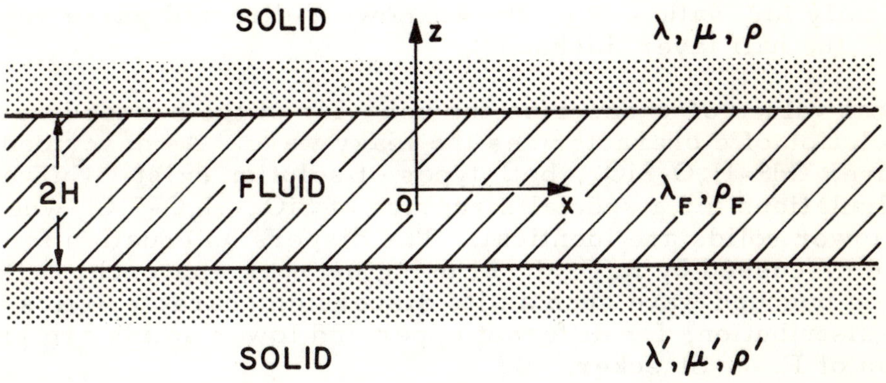

Fig. 8 - Layered Geometry: Ideal fluid bounded by isotropic semi-infinite solids.

The dispersion relation is found to be in the form of Eq. (1).

$$\left[G - \cot kH\sqrt{\left(\frac{c}{c_F}\right)^2 - 1}\right]\left[G + \tan kH\sqrt{\left(\frac{c}{c_F}\right)^2 - 1}\right] = (G - F) \cdot \left[G - \frac{1}{2}\left(\cot kH\sqrt{\left(\frac{c}{c_F}\right)^2 - 1} - \tan kH\sqrt{\left(\frac{c}{c_F}\right)^2 - 1}\right)\right]$$

where

$$F \equiv \frac{\rho_s}{\rho_F}\left(\frac{c_t}{c}\right)^4 \sqrt{\left(\frac{c}{c_F}\right)^2 - 1}\left[\frac{\left(2 - \left(\frac{c}{c_t}\right)^2\right)^2}{\sqrt{1 - \left(\frac{c}{c_p}\right)^2}} - 4\sqrt{1 - \left(\frac{c}{c_t}\right)^2}\right] \qquad (1)$$

$$G \equiv \frac{\rho'_s}{\rho_F}\left(\frac{c'_t}{c}\right)^4 \sqrt{\left(\frac{c}{c_F}\right)^2 - 1}\left[\frac{\left(2 - \left(\frac{c}{c'_t}\right)^2\right)^2}{\sqrt{1 - \left(\frac{c}{c'_p}\right)^2}} - 4\sqrt{1 - \left(\frac{c}{c'_t}\right)^2}\right]$$

c_p and c_t are the velocities of longitudinal and shear waves, respectively, in the solid, and can be expressed in terms of the basis physical constants as

$$c_p^2 = \frac{\lambda_s + 2\mu_s}{\rho_s} \quad ; \quad c_t^2 = \frac{\mu_s}{\rho_s}$$

Primed quantities characterize the lower solid. $c_F^2 = \lambda_F / \rho_F$ is the sonic velocity in the fluid. c and k are the phase velocity and wave number in the layered system propagating in the x-direction.

To investigate the dispersion characteristics of this structure, we solve (1) numerically for values of c, the unknown x-directed phase velocity, in terms of H, the half layer thickness.

The real values of c describe the behavior of the bounded wave and the complex solution of c characterizes the leaky wave. As described earlier for the structure, $CdS\text{-}H_2O\text{-}SiO_2$, both types of solution exist. Curves in Fig. 7 are the calculations for a special structure of $SiO_2\text{-}H_2O\text{-}SiO_2$ where both the upper and lower solids are identical. The dispersion equation for identical solids were discussed by R. C. M. Li using network approach. Detailed information on bounded wave such as its dispersion characteristics as well as its energy distribution, for different upper and lower solids are given in the dissertation of P. W. Staecker.

IV. Applications:

Based on the use of a thin fluid (or grease) layer sandwiched between a crystal plate and a surface wave delay line, three devices have been successfully realized; a surface wave amplifier; a delay line of short length which produces large delays and a coupler which couples surface waves between substrates (the substrates can be non-piezoelectric).

(A) Amplification: The amplifier configuration is shown in Fig. 9. The optical axis of the CdS crystal is perpendicular to the major surface of the quartz plate. The output pulse which has been delayed due to the presence of liquid layer and CdS plate is piezoelectrically coupled to the conduction electrons. Since the CdS crystal is a photoconductor, the output pulse is highly photosensitive. When a sufficient d. c. voltage is applied to the CdS plate with polarity shown in Fig. 9. The output pulse is then amplified. By adjusting

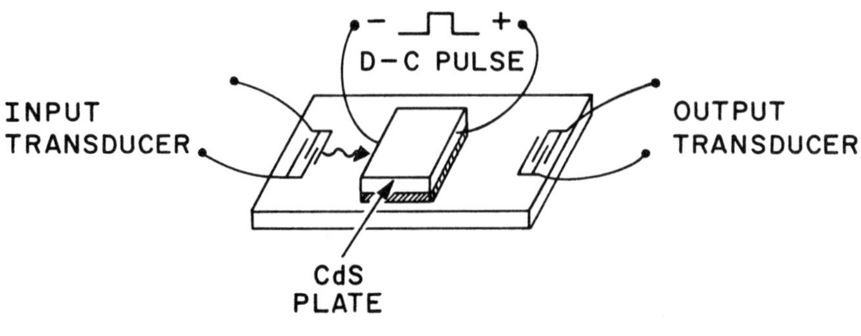

Fig. 9 - Experimental configuration for amplifying surface waves.

the layer thickness of the fluid layer, we have been able to amplify either the bound wave or the leaky wave. The leaky wave amplification (thin layer) is essentially the amplification of shear waves in CdS which can be very efficient. In fact, a net gain more than 20 dB and an electronic gain of 60 dB have been observed at 100 MC pulsed signal with a field intensity 2000 volts/cm, the structure for which is of a thin grease layer sandwiched between a CdS plate (ℓ: 0.6 cm, h: 0.2 cm) and a $LiNbO_3$ surface wave substrate. And also an electronic gain of 75 dB is observed at 160 MC for the structure of CdS-liquid-quartz. The amplification of bounded wave is also found quite efficient which is due to the fact that there is deep penetration of elastic energy into the CdS plate (Staecker dissertation).

(B) Long Time Delay: Picture 10a shows both the feed-through input signal and the output signal of a quartz surface wave delay line operated at 160 MC. (The surface of the delay line is well polished and free of any visible scatterer.) Picture 10b shows the output signal of the same delay line but with the introduction of a thin liquid layer and a polished CdS crystal (1 cm^3). An additional time delay of $\approx 23\ \mu sec.$ is seen in picture 10b. The surprising finding here is that with the introduction of a layer structure one actually observes an increase in signal amplitude in addition to gaining longer time delay. This can be interpreted by that the attenuation of surface wave at high

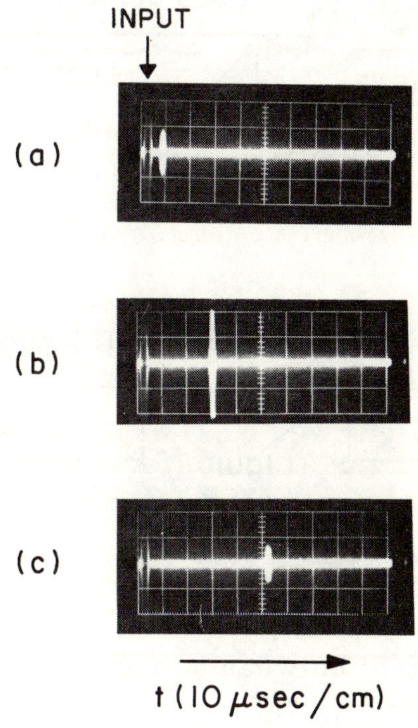

Fig. 10 - Long Time Delay:
(a) Signal observed at the output transducer of the quartz Rayleigh wave delay line substrate. The delay line is operated at 160 MC and its surface is free of liquid - CdS overlay. The first pulse corresponds the input signal radiating through the air.
(b) Output response of the structure with liquid - CdS overlay.
(c) Same condition as (b) except that the CdS crystal is placed at a different angle.

frequencies is very high in comparison with that of a bulk wave due to the scattering of surface imperfections and air loading, etc.. Picture 10c is taken under the same condition as 10b except that the CdS crystal is now placed at an angle different from that for picture 10b. An additional time delay of ≈ 23 μsec is obtained in this case. The mechanism of long time delay results from multiple reflections of sonic waves inside the CdS crystal. Multiple reflections clearly, can be controlled by the crystal dimensions, however, it can also be caused by the unevenness of the fluid layer or due to both. Supposing the layer velocity at thickness H_1 is $c_L(H_1)$, the direction of propagation of the leaky shear wave with respect to the plate normal would be $\theta = \sin^{-1}(c_t/c_L(H_1))$, after the shear wave reflected back from the top boundary, it will reconvert into a surface wave efficiently at the liquid interface only if the layer thickness is still in the neighborhood of H_1. If it does not, the wave will bounce back from the liquid interface and continue the process of bouncing until the phase matching condition is satisfied and the wave reconverts into the surface wave.

(C) Coupler: The fluid layer scheme shown in Fig. 11a has been proven to be an efficient means to couple acoustic surface wave from one substrate to the other. A coupling loss of less than 3 dB has been observed with thin

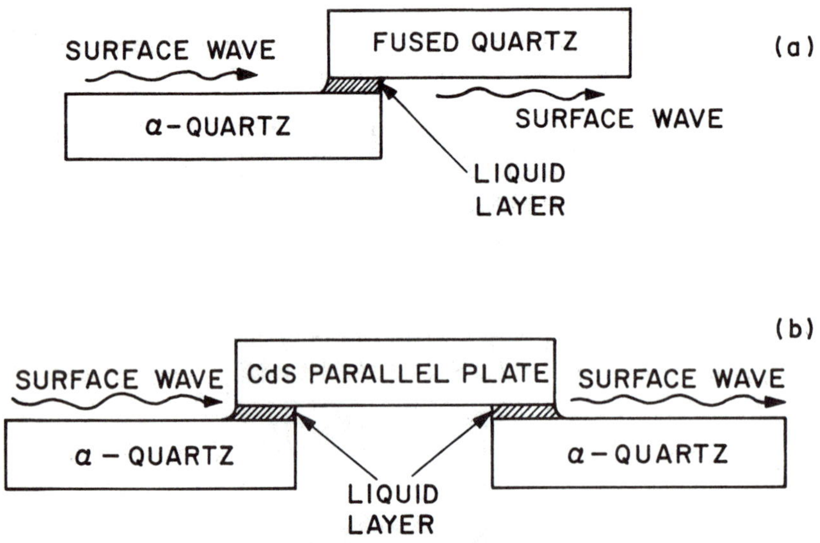

Fig. 11 - Surface wave coupling between substrates using a liquid layer.

water layer as coupling agent between two quartz plate. The operating frequency is 30 MC. It is understood that for surface wave to surface wave coupling the layer wave is required to be bounded.

Leaky layer wave can be also used as a coupling scheme between surface waves by adopting the structure of Fig. 11b.

Acknowledgment

The author wishes to thank P. Staecker, R. C. M. Li and I. Drukier for their contributions to the theory and experiment.

References

1. W. C. Wang, P. Staecker, and R. C. M. Li, Appl. Phys. Lett., 16, 291 (1970). Also IEEE Ultrasonic Symposium, St. Louis, Mo. (1969). Erratum, Appl. Phys. Lett. 18, p. 66 (1971).

2. W. C. Wang, Jour. Appl. Phys., 41, 1 (1970).

3. A. A. Oliner, R. C. M. Li, and H. L. Bertoni, PIB Res. Report PIBEP-69-053.
   K. H. Yen and R. C. M. Li, PIB JSTAC Report No. R-452. 35-70, p. 26 (1970).

4. P. Lloyd and M. Redwood, Acoustica, 16, 224 (1965/66).

5. J. Wilkinson and C. D. W. Wilkinson, Elec. Letters, 6, 426 (1970).

6. P. Das and M. N. Araghi, Appl. Phys. Letters, 16, 293 (1970).

7. A. P. Van Dan Heuvel, D. B. Owen, and S. G. Joshi, IEEE Ultrasonic Symposium, St. Louis, Mo. (1969).

8. F. Y. Cho, B. J. Hunsinger, and R. L. Lawson, Appl. Phys. Letters, 16, 441 (1970).

9. P. Staecker, Dissertation (June 1970), EE Dept., Polytechnic Institute of Brooklyn.

ACOUSTIC SURFACE WAVE ATTENUATION STUDIES IN SUPERCONDUCTING THIN FILM SYSTEMS

Dr. Mahesh C. Jain
Department of Physics
University of Essex
Colchester, Essex, U.K.

INTRODUCTION

Ultrasonic attenuation studies in bulk single crystals and polycrystalline materials have been carried out by numerous scientists studying various effects and phase transitions. To study these effects in thin films was not easy until after the generation of elastic waves at the surface of piezoelectric solids was reported (1-5). Generation of surface waves and their properties have been discussed in detail by several scientists (1-5). Of these methods 1, interdigital electrodes (3) and 2, a simple metallic grating with all its lines connected at their ends(5) are very well suited for work in vacuum and at low temperatures.

Most of the physical properties of a metal are determined by free electrons and lattice vibrations and the interaction between the two. The transport properties give indirect information about the interaction, while in case of ultrasonic attenuation,measurements yield directly the interaction between the lattice waves and the conduction electrons at low temperatures. The cause of attenuation is attributed to the electron-phonon and phonon-phonon interaction, the later being present in insulators as no free electrons are available. In metals electron-phonon interaction is predominent(6).

So the study of ultrasonic surface wave attenuation in thin films is an important technique in studying the interaction between electrons and phonons. Very recently surface waves have been used to study superconducting transitions of first and second order (7-9) and the proximity effect (10,11) in thin films. The main advantage of this technique over other techniques is that the attenuation depends on the behaviour of the composite specimen as a whole. The contribution to the attenuation by the edges is negligible as they are thin and of very very small area in proportion to the whole film, so one does not beed to trim the edges of the film. In this paper we shall discuss the experimental aspects of the surface wave attenuation in thin films of superconductors and proximity effect specimens. Some results obtained by the author and other research workers on such films will also be mentioned.

Methods of exciting surface waves.

Of the two methods to excite surface waves (3,5) mentioned earlier, the interdigital electrodes (3) deposited on to the surface of piezoelectric substrate are widely used for most of the applied work. In these electrodes rf voltage is applied between interleaved comb like electrodes to excite surface waves. These electrodes are deposited directly on to the surface of piezoelectric substrate by photoetching of a thin metal film(3). The wavelength of the excited surface waves by such spatially periodic electrodes is equal to the period. These electrodes have been successfully used to study attenuation in thin film superconductors (8, 10). The only limitation of this method is as one goes high in frequency is the fact that the spacing between the leaves of the electrodes decreases and one needs special technique and apparatus to deposit these interdigital electrodes with spacing below 10 .

We used the second method (5) in which two metallic gratings (one generator and other detector), with grating spacing of 10μ, were deposited parallel to each other on a piezoelectric substrate. The frequency of operation is twice in this case for the same periodicity as would have been achieved by interdigital electrodes. Fig. 1 shows the arrangement to excite surface waves at 316MH for a 10μ grating spacing. In this method the grating spacing equals the wavelength at frequency of operation. The metallic gratings were produced by photoetching a thin ($\sim$ 1000Å) chromium film on a y- cut quantz substrate with surface wave travelling along x-axis. For this crystal, the directions of phase velocity and energy transport are colinear. In x-cut quantz plate, the direction of the phase velocity (which is perpendicular to the grating lines) is parallel to Y-axis, but the elastic energy is transported at an angle of roughly 8° relative to the phase velocity[12]. Shipley AZ-1350 photoresist and a master 10μ grating were used; acidic $Ce(SO_4)_2$ solution was used for etching. The gratings made from chromium last for a long time. To study ultrasonic surface wave attenuation in comparatively thin films (500 $\sim$1000 Å thick) and to notice appreciable measurable change one needs to use the high frequencies. This method (5) is simple and easy and has been used very successfully (7,9,11)

Fig. 1 Arrangement for excitation of elastic surface waves at 316 MHz on Y-cut quantz bar (1 cm x 5cm x.1cm), distance between gratings about 2 cms, 10μ grating spacing, about 400 lines on each grating [After (5)]

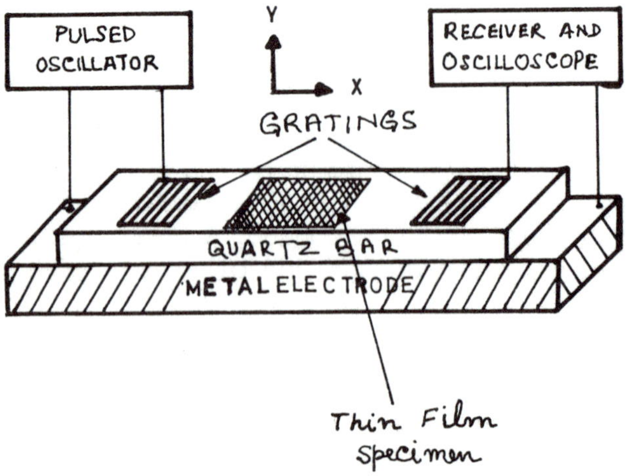

Preparation of thin film specimen:

Thin films were deposited on a quantz substrate (Fig 1 between the two parallel gratings) in a conventional evaporator. The evaporator was equipped with a liquid N_2 cooled Miessner trap, crystal thickness monitor and a mask changing device. An ultimate pressure of 2×10^{-7} Torr was reached and the pressure during evaporation was less than 3×10^{-6} torr.

As the acoustic surface wave attenuation in thin films due to free electrons depends on the electron mean free path(m.f.p.) precautions were taken to deposit relatively clean films with long m.f.p. This is done by taking 1, a pure starting evaporating material (~6N purity), 2, good vacuum and 3, as fast as possible a rate of evaporation to minimise the contamination from the residual gases in the vacuum chamber. A slow rate of evaporation does not help in securing a film with longer m.f.p. though the vacuum during the evaporation is slightly improved. A liquid N_2 cooled substrate also does not help very much, though in single film measurements we find that it gives planer films but short m.p.f.

For proximity effect specimens we used triple layer Cu-Pb-Cu sandwiches. There was an internal of only few seconds between the evaporation of successive film to avoid the oxide film formation between these films. Cu-Pb system was chosen because of the low solubility of Cu in Pb[13]. As far as possible, the two Cu films were made of the same thickness d_{cu} because the theory is much easier in this case. In fact for d_{cu} greater than about 1000Å the critical temperature is a very slow function of d_{cu}, so a small change between the thicknesses of two Cu films in a specimen is not very significant.

Monitor strips of the metal on separate glass slides were deposited at the same time. From these the thicknesses of the films were optically checked and the m.f.p. was estimated from the conductivity measurements. A separate Pb monitor strip was also deposited at the same time on a quartz substrate (with surface wave electrodes) to estimate the changes occuring in the proximity effect specimen from that of the pure lead.

Low Temperature measurements and Results:

The specimen was placed in the inner chamber of a double chamber vessel which was then placed in a conventional croyostat. The pulse height was integrated using Brookdeal Box-car (Lineargate 415, scan 425A) unit and a X-Y recorder. Temperatures above 4.2 were obtained by heating the inner chamber with helium as exchange gas. Thc heater wire was wound over the whole length of the inner chamber to make temperature gradient across the film to a minimum. Temperatures below 4.2 were reached by pumping over the helium bath. The temperature was measured by a calibrated germanium resistance thermometer using an A.C. Bridge. The stability of temperature in this way was found to be very good over a long period during which measurements were carried out at one particular temperature. Fig 2 displays the scan of the oscillogram of the received pulse at 1.33°k after going through a Cu - Pb - Cu sandwich.

The attenuation measurements were made by measuring the change in attenuation from superconducting to normal state by applying a magnetic field or by noticing the change in attenuation below the critical temperature. It was noted that no change in normal state attenuation appeared during cooling to below 1.2°k from the critical temperature T_c. For magnetic fields less than about 200 Gauss we used a calibrated air cored Helmholtz pair and for higher fields we used an iron cored electromagnet. We aligned the magnetic field relative to the specimen with the aid of the sharp minimum in attenuation in parallel field.

Fig. 2 Actual scan of the oscillogram of the received signal after being attenuated by a Cu-Pb-Cu sandwich.

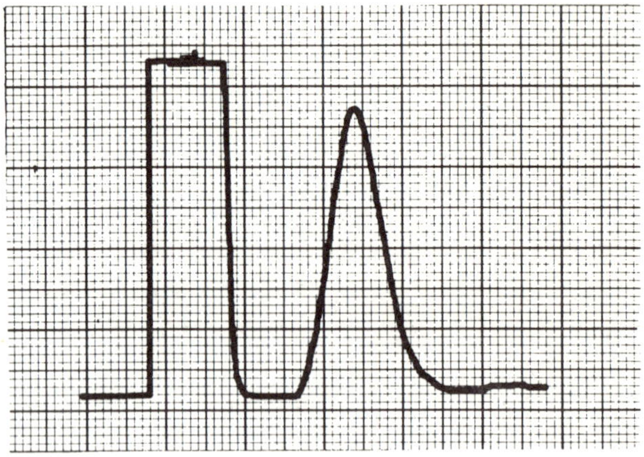

Figs. 3a,b show the attenuation change in a, Pb film (2680Å) at 1.48°k and b, Cu (1070Å) - Pb (2680Å) - Cu (1070Å) sandwich at 1.33°k. The Pb film was deposited at the same time as Cu-Pb-Cu on a separate quartz substrate with surface wave electrodes. The changes occuring are obvious. We shall not discuss here the theoretical aspects of the subject. From such measurements we have studied the following things which are discussed in detail in the respective reference quoted.

1. <u>Parallel field Case</u>

a. In parallel field deduction of the pair potential field dependence shows agreement with an approximate theory (9)

b. At present we are examining the critical parallel field in a proximity effect specimens. (To be published).

2. <u>Perpendicular field</u>

a. In perpendicular field the attenuation near Hc_2 varies in the way theory predicts for bulk behaviour in the dirty limit (9).

b. A study of perpendicular critical field versus temperature of various Cu-Pb-Cu sandwiches (11)

The other things which have been so far reported by other research workers on superconducting thin films using acoustic surface waves include attenuation in thin films (7), superconducting phase transitions (8) and the proximity effect (10).

a b

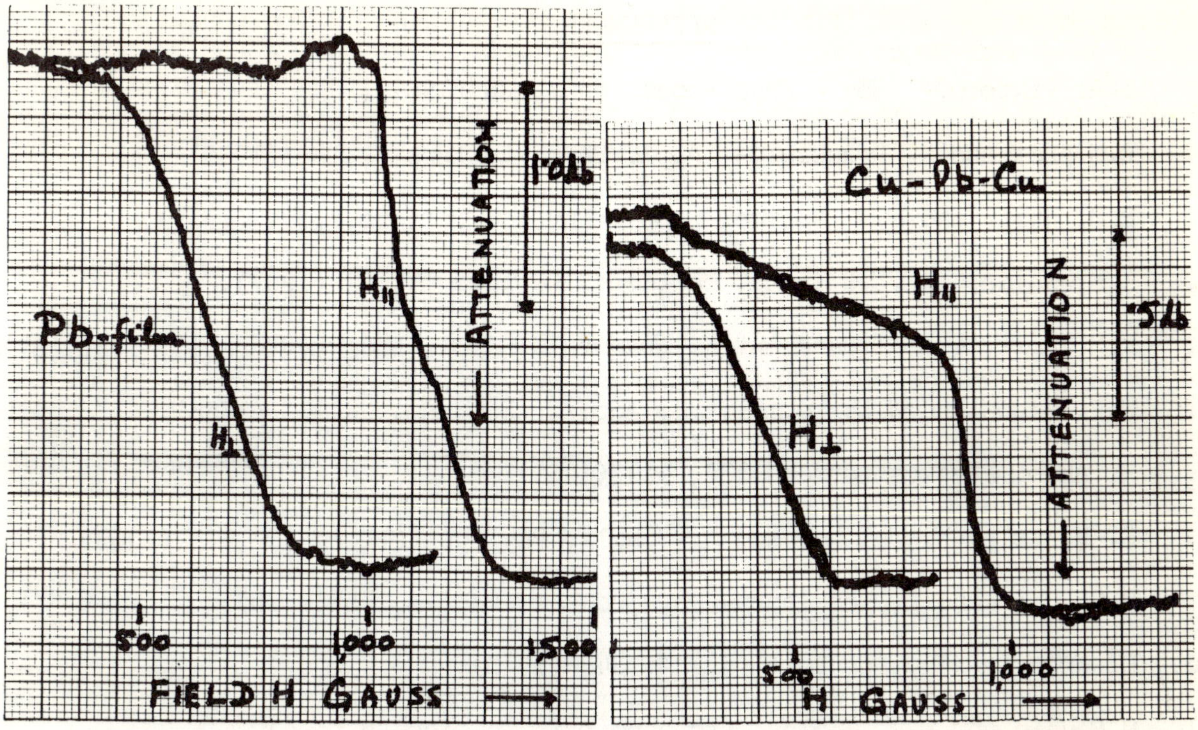

Fig 3. Actual recording of the change in ultrasonic surface wave attenuation versus parallel ($H_{\parallel}$) and perpendicular ($H_{\perp}$) magnetic fields (a) in Pb (2680Å) film at 1.48°K, (b) in Cu (1070Å) - Pb (2680Å) - Cu (1070Å) sandwich at 1.33°K. Pb film and sandwich were deposited at the same time.

The technique using acoustic surface waves to study thin films is a new method to study the electron-phonon interaction and because this method is still in developing stage we think it will be used to study many more properties of thin films besides that of superconductors,e.g. magnetic phase transitions etc.

REFERENCES

1. I.A. Viktorov- Sov - Phys Acoustics 7,236 (1962).

2. R.M.Arzt and K. Dransfeld- Appl. Phys. Letters 7, 156 (1965)

3. R.M. White and F.W. Voltmer- Appl. Physics Letters,7,314 (1965).

4. R.M. White- I.E.E.E Trans. on Elect. Devices 14,181 (1967).

5. R.M. Arzt, E. Salzmann and K. Dransfeld- Appl. Phys Letters 10, 165(1967)

6. W.P. Mason and H.E. Bommel; J. Acoust. Soc. Am. 28,930 (1956)

7. F. Akao; Physics Letters 30A, 409,(1969).

8. E. Kratzig, K. Walther and W. Schilz; Physics Letters 30A,411 (1969).

9. M.C. Jain and L. Mackinnon- Physics Letters 32A,275 (1970).

10. E. Kratzig- Physics Letters 33A, 343 (1970).

11. M.C.Jain and D.R. Tilley- (to appear in Physics Letters)

12. C.A. Coquin and H.F. Tiersten; J. Acoust. Soc. Am. 41, 921 (1967).

13. M. Hansen- Constitution of Binary Alloys (McGraw-Hill Book Co., N.Y. 1958).

ACOUSTIC SURFACE-WAVE TRANSVERSAL FILTERS

Carlo Atzeni, Leonardo Masotti
Istituto di Ricerca sulle Onde
Elettromagnetiche del Consiglio
Nazionale delle Ricerche.
Firenze, Italy

ABSTRACT.

Acoustic surface-wave tapping by interdigital arrays of transducers offers a new powerful tool for the implementaion of linear transversal filters. The design of the interdigital configuration for the synthesis of the filter's impulse response can be properly understood from the theory of the particular sampling procedure outlined in this paper.

Examples of the application of this design criterion to the synthesis of matched filters of various frequency responses are reported.

1. - INTRODUCTION.

The chief and completely new feature of acoustic surface-wave delay lines consists in the possibility of a direct access to the acoustic signal in any point of the propagation path. Surface-wave tapping appears, therefore, to offer a powerful tool for the implementation of linear transversal filters meeting the signal processing needs of today's radar and communication systems.

In linear transversal filters the required impulse response is synthesized in a sampled form by means of a tapped delay line with weighted and summed taps. The synthesis procedure consists of sampling the impulse response at appropriate intervals and using the sample values as the tap weights ([1]).

The sampling procedure used in acoustic surface-wave transversal filters consists in taking the samples of the specified impulse response at points where the instantaneous phase of the response is multiple of π radians ([2]). Transversal filters synthesized according to this particular sampling criterion, referred to as Phase-Sampled Impulse Response (PSIR) filters, do not require phase adjustments of taps, apart from phase shifts of π.

In PSIR acoustic surface-wave filters surface-wave tapping is simply achieved by using a two-terminal "interdigital" array of conductive electrodes, deposited on a piezoelectric substrate, with the electrodes suitably connected to two output terminals to obtain phase inversions. Arbitrary response functions can be synthesized by means of an interdigital configuration with the number, spacing and shape of the tapping electrodes determined by the mentioned sampling procedure. If the response function has a constant carrier frequency, the electrodes are spaced of half the corresponding acoustic wavelength; if the response function is frequency-modulated, the electrodes are nonuniformly spaced, and their separations are graded in accordance to the frequency modulation law.

In both cases, if f_o is the mean frequency and T the duration of the waveform, an interdigital array of $2f_oT$ electrodes is needed. For waveforms of low fractional bandwidth, however, this number can be reduced, in principle, down to a minimum number which depends on the value of the fractional bandwidth.

In this paper the basic principles of the sampling procedure for PSIR filter synthesis are reported. As an example, their application to the design of linear frequency modulated and linear period modulated matched filters is described, and experimental results are given.

2. - SAMPLING PROCEDURE FOR PSIR FILTER SYNTHESIS.

a - Sampling criterion.

Let the filter impulse response be represented by the amplitude and frequency modulated function (fig. 1a):

$$(1) \quad h(t) = a(t)\cos\left[2\pi f_o t + \theta(t)\right] = a(t)\cos\phi(t) \qquad |t| \leq \frac{T}{2}$$

where f_o is the mean value of the instantaneous frequency;

$f(t) = f_o + \frac{1}{2\pi}\frac{d\theta(t)}{dt}$. This waveform is assumed to be narrow-band, with a bandwidth B centered on the frequency f_o.

Let us consider that the function (1) is sampled in a suitable set of points t_n. The function, reconstructed from its sampled version by filtering the band B around f_o, has the form:

$$(2) \quad h(t) = \sum_{-\infty}^{+\infty}{}_n\, a(t_n)\cos\left[2\pi f_o(t-t_n) + \phi(t_n)\right]\operatorname{sinc}\left[B(t-t_n)\right]$$

According to the Shannon theorem, the sampling interval is $t_n - t_{n-1} = 1/B$, so that the number of the sampling points in the time T is N = B T. For each sampling point, however, both the amplitude, $a(t_n)$, and the phase, $\phi(t_n)$, of the function must be specified.

The basic principle of the sampling procedure used in PSIR filter synthesis consists in choosing as sampling points the times t_n where the instantaneous phase, $\phi(t)$, is multiple of π, i.e.:

$$(3) \quad \phi(t_n) = n\pi \qquad \text{(n integer)}$$

The resulting sampling intervals, $t_n - t_{n-1}$, are nonuniform in the time domain and depend on the phase modulation law.

The reconstructed waveform assumes now the form:

$$(4) \quad h(t) = \sum_{-\infty}^{\infty}{}_n\, a(t_n)(-1)^n \cos\left[2\pi f_o(t-t_n)\right] \mathrm{sinc}\left[B(t-t_n)\right]$$

The fundamental difference with respect to the form (2) is that, according to (3), the phase of the function in the sampling points is always a multiple of π, thus involving merely inversions in sign of the amplitude samples.
The features of this sampling procedure, i.e. limits of applicability, spectral distribution of the sampled function and number of samples, are now briefly reviewed. A more extensive treatment can be found in the original paper (2).

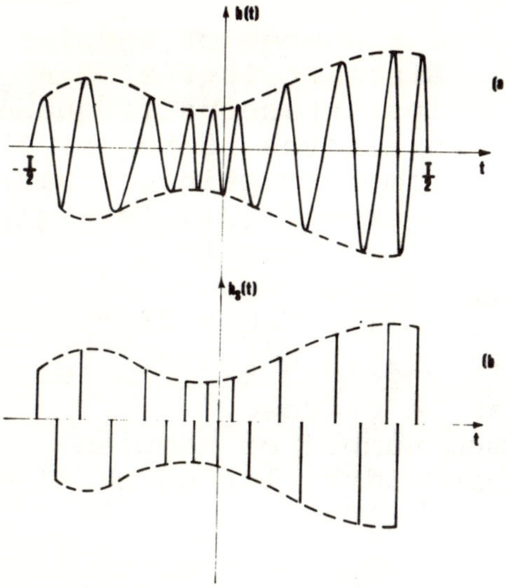

Fig. 1 - a) AM FM waveform, h(t); b) Succession of samples of h(t). The sampling points occur where the instantaneous phase of h(t) is multiple of π radians.

b - Spectral distribution. It can be shown that the sampled version $h_s(t)$ of h(t) (fig. 1b) can be written as:

$$(5) \quad h_s(t) = \sum_{\infty}^{\infty}{}_n\, a(t)(-1)^n\, \delta(t-t_n) =$$

$$= \sum_{1}^{\infty}{}_{m\ \mathrm{odd}}\, a(t) \cos m\phi(t)$$

The fundamental term of this series is the original function (1) and hence its spectrum coincides with the spectrum H(f) of h(t).

The general term with $m \neq 1$ has the same envelope a(t) of h(t) and an instantaneous phase $m\phi(t)$. Using the principle of stationary phase, it can be shown that its spectrum, $H_m(f)$, has an amplitude at the frequency f which is proportional to that of H(f) at the frequency f/m. It follows that the bandwidth B_m of the m-th term of the series (5) is m times that of h(t), i.e. $B_m = mB$, and is centered on $f_{om} = mf_o$, so that the fractional bandwidth, B_m/f_{om}, is constant for any m.

The spectral distribution which ensues is shown in fig. 2. The original waveform h(t) can be reconstructed from its sampled version $h_s(t)$ provided that the fundamental spectrum (m=1) is not overlapped by any spectrum with $m \neq 1$. Since all the even harmonic terms are missing, it can be recognized that this condition is met if

$$(6) \quad B \leq f_o$$

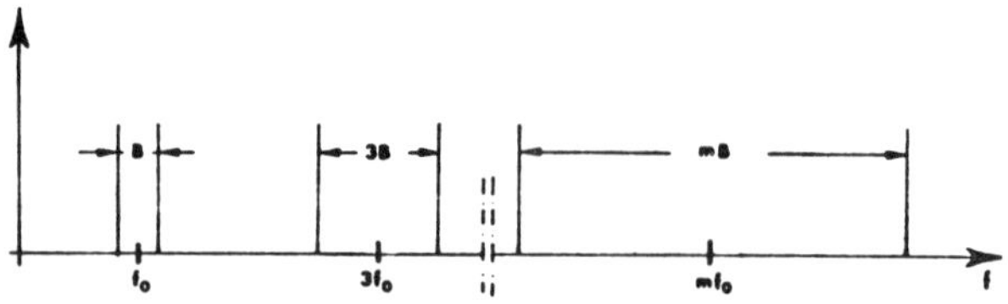

Fig. 2 - Spectral band distribution of the sampled waveform. The fundamental spectrum is that of the original waveform.

Thus, this particular sampling procedure can be applied only if the fractional bandwidth B/f_o of h(t) is less than unity.

c - Number of samples.
Since f_o is the mean instantaneous frequency of h(t), the number of the sampling points t_n, derived from eq. (3), is given by:

$$(7) \quad N = 2f_oT$$

However, if the signal fractional bandwidth B/f_o is less than 50%, this number is redundant and can be reduced.

In order to determine the minimum number of samples consistent with a given fractional bandwidth, let us consider the function:

$$(8) \quad h^k(t) = a(t) \cos \frac{\phi(t)}{k}$$

derived from h(t) by scaling down the instantaneous phase $\phi(t)$ of an odd integer k, thus having a mean frequency f_o/k and a bandwidth B/k.

By applying the sampling procedure above described, a number of:

$$(9) \quad N_k = 2 \frac{f_o}{k} T$$

sampling points is determined, occurring at the times t_n^k where:

$$(10) \quad \frac{\phi(t)}{k} = n\pi$$

and thus coinciding with one every k of the sampling points given by eq. (3).

By a procedure parallel to that used in the previous section, it can be shown that the spectral distribution of the N_k samples of function (8) consists of a series of harmonic spectra of constant fractional bandwidth, B/f_o, with a fundamental band of width B/k around the frequency f_o/k (fig. 3). The k-th spectrum of this distribution coincides with the spectrum H(f) of h(t). Thus, the reconstruction of h(t) by band pass filtering requires that the k-th spectrum is not overlapped by any spectrum $m \neq k$.

It is recognized that this condition is met if:

$$(11) \quad \frac{B}{f_o} \leqslant \frac{2}{k+1}$$

The maximum value k_M of k is the maximum odd integer consistent with condition (11). Assuming for the sake of simplicity that f_o/B is an integer, one obtains:

$$(12) \quad k_M = 2 \frac{f_o}{B} - 1$$

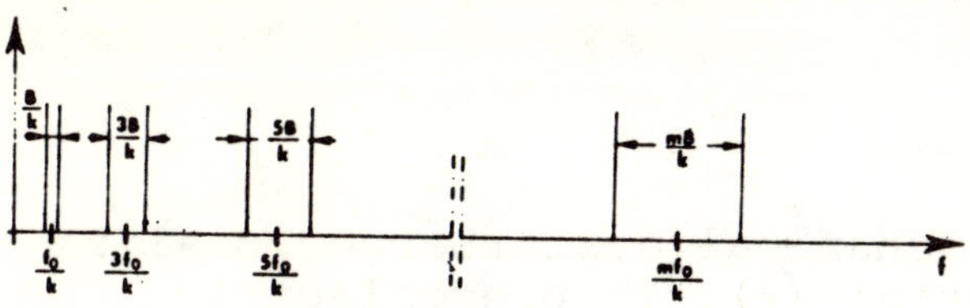

Fig. 3 - Reduction of the sample number: spectral band distribution. The fundamental spectrum is that of a waveform of bandwidth B/k and center frequency f_o/k. The k-th spectrum is that of the original waveform.

Thus, from (9), the minimum number of sampling points results:

$$N_{min} = BT \frac{1}{1 - \frac{B}{2f_o}} \qquad (13)$$

For very low fractional bandwidths ($B/f_o \ll 1$) it appears that $N_{min} \simeq BT$.

3. - PSIR TRANSVERSAL FILTERS.

A PSIR transversal filter consists of a tapped delay line, with the tap configuration determined by the above sampling law (fig. 4).

The distance between two adjacent taps is given by:

$$X_n - X_{n-1} = v(t_n - t_{n-1}) \qquad (14)$$

where v is the propagation velocity in the delay line, and the delays t_n are derived from eq. (3).

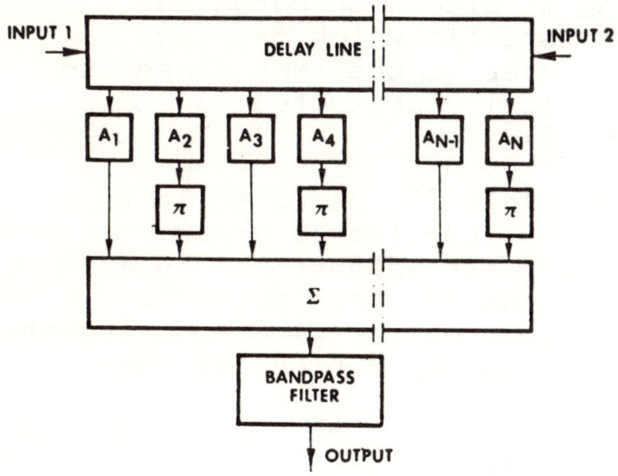

Fig. 4 - PSIR transversal filter.

The taps must be weighted according to the amplitude samples, $a(t_n)$, and must be suitably connected to the output terminals to obtain phase inversions. The signal to be processed propagates along the delay line, where the taps, which are assumed not perturbing the signal propagation, spatially sample the travelling wave. The output is formed by summing the contributions from the taps, and must be band-pass filtered in order to select the band of interest. In fact, as shown above, the transfer function of the unfiltered device contains all the odd harmonics of the desired response. The synthesis of PSIR transversal filters is particularly simple when the required impulse response, h(t), has a rectangular envelope. In this case in fact, the samples coincide with maxima or minima of h(t), so that $a(t_n) = 1$ for any n.

The taps of the delay line, therefore, must have all the same weight.

4. - ACOUSTIC SURFACE-WAVE FILTERS.

An acoustic surface wave is a mechanical disturbance localized on the surface of an elastic material (3). In piezoelectric crystals the particle motion is associated with an electric field which extends into the medium as well as outside the medium with a decay rate of an acoustic wavelength (4).

Several types of transducers have been devised for generation and detection of surface waves (5). The most simple and efficient transducer consists of an array of conductive electrodes deposited on the surface of the piezoelectric substrate. When driving the array with an external voltage, a spatially nonuniform time-varying electric field is generated, which produces local stresses in the medium and launches elastic surface waves. The transducer operation is reversible, i.e. a time-varying output voltage results from the contributions of the instantaneous electric fields associated with the elastic wave propagating under the electrodes. The output signal, which is constructed by adding spatial samples of the travelling acoustic waveform, depends on the tapping electrodes configuration.

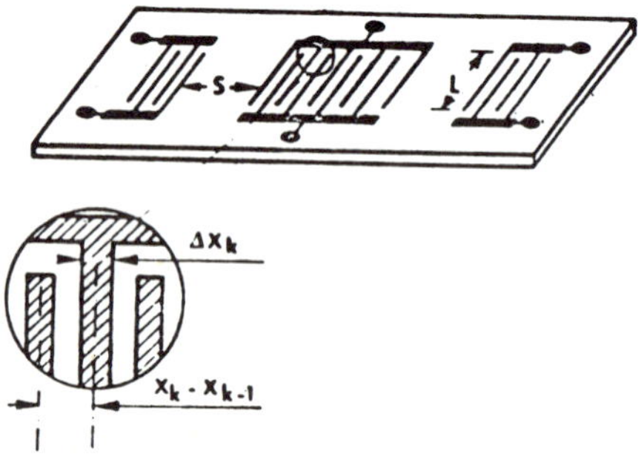

Fig. 5 - Interdigital structure of acoustic filter.

According to the general scheme of transversal filter described in the previous section, the frequency response of the transducer configuration is determined by the number, spacing and shape of the electrodes. Phase inversions of the taps are simply achieved by using an alternate-potential structure, in which tapping electrodes of the same phase are electrically connected together (fig. 5).

This structure is referred to as "interdigital", since it looks like two interleaved combs.

Finally, tap weights can be introduced by varying the relative lengths of the electrodes, since the piezoelectric coupling depends on the electrode length.

Interdigital masks can be easily drawn and reproduced onto the optically polished surface of a piezoelectric crystal by standard photolithographic processes: the fabrication techniques are the same used for integrated circuits.

Thus far no consideration was given to the method of transduction by which the acoustic wave is launched. Ideally, a transducer is needed which does not introduce distorsions of the signal to be processed. In practice this condition can be well approximated by using an input transducer which operates over a band large enough compared with that of the signal. The transmitter transducer consists of M pairs of interdigital fingers of constant spatial frequency v/f_o, whose impulse response is a rectangular pulse of constant carrier frequency f_o and of duration M/f_o. The transmitter transfer function is therefore given by:

$$T(f) = \frac{\sin\left(M\pi \frac{f-f_o}{f_o}\right)}{M\pi \frac{f-f_o}{f_o}} \qquad (15)$$

Thus, in order to have a transmitter bandwidth, f_o/M, large in comparison with the signal bandwidth, B, the number of the finger pairs of the transmitter array must be suitably less than the signal inverse fractional bandwidth, f_o/B. In an interdigital structure this number cannot be less than $M = 1$.

The transmitted acoustic wave propagates along the surface in both directions, the two acoustic powers being equal by symmetry. The wave propagating in the undesired direction must be terminated by an acoustic absorber.

5. - AN APPLICATION: SYNTHESIS OF MATCHED FILTERS.

One of the most important problems in modern communication systems is the synthesis of matched filters for the extraction of signals from noise [6].

A filter is "matched" to a signal s(t) of spectrum S(f), when its transfer function is $H(f) = S^*(f)$. Alternatively, in the time domain the impulse response of matched filter is required to be:

$$h(t) = \tilde{s}(-t) \qquad (16)$$

i.e. the time-inverse of the signal to be processed.

The technique of transversal filtering is particularly suited for the implementation of matched filters, since one and the same filter can be used for both signal generation and signal processing [1,2].

In fact, with reference to fig. 2, if $h(t) = s(t)$ is the impulse response of the tapped delay line when driven from the input 1, the impulse response becomes $h(-t) = s(-t)$ when the same delay line is driven from the input 2.

In this section we give the design criterion for the synthesis of PSIR acoustic surface-wave matched-filters for two types of signals of interest in radar applications.

a - Linear FM signal.

A rectangular-envelope linear FM signal is described by the function:

$$\tilde{s}(t) = \mathrm{rect}\left(\frac{t}{T}\right) \cos\left[2\pi\left(f_o t + \frac{\alpha}{2} t^2\right)\right] \qquad (17)$$

where α is the rate of the linear frequency modulation:

$$f(t) = f_o + \alpha t$$

According to eq. (3), the sampling points t_n are given by the solution of the equation:

$$2\pi \ (f_o t + \frac{\alpha}{2} t^2) = n\pi$$

which provides:

$$(18) \qquad t_n = - \frac{f_o}{\alpha} + \sqrt{(\frac{f_o}{\alpha})^2 + \frac{n}{\alpha}} \qquad |t_n| \leq T/2$$

Since the function (17) has a rectangular envelope, its samples at the times (18) coincide, as mentioned above, with the successive maxima and minima of the function itself, thus being all equal in amplitude and alternate in sign.

If the fractional bandwidth of s(t) is less than 50%, the number of samples can be reduced. For a reduction factor k, samples must be taken, from eq. (10), at the times:

$$(19) \qquad t_n^k = - \frac{f_o}{\alpha} + \sqrt{(\frac{f_o}{\alpha})^2 + \frac{kn}{\alpha}}$$

An experimental model of acoustic surface-wave matched filter has been built (7) for a linear FM signal having the following characteristics:

center frequency	f_o = 9Mhz
bandwidth	B = 2,5Mhz
duration	T = 32 μsec

The number of taps (7) is N = 576.

In fig. 6a and b the experimental results obtained using an interdigital array of 576 copper electrodes deposited on an x-cut y-propagating quartz plate are shown. The same delay line was used to generate the signal (top of fig. 6a) and to correlate the so passive-generated pulse (bottom of fig 6a and 6b) the measured time-delay vs. frequency. The characteristic is shown in fig. 7.

Since the fractional bandwidth of the signal is less than 50%, the number of taps of the interdigital array can be reduced. From (12) the maximum reduction factor results $k_M = 5$.

In fig. 8 the output signal from a line having the minimum number of taps is shown.

The minimization of the number of electrodes in an interdigital configuration for matched filtering of linear FM signals has been extensively discussed in a previous paper (8).

b - Linear period-modulated signal.

A linear period modulated (i.e. hyperbolic frequency modulated signal (9) of rectangular envelope is described by the function:

(20) $$g(t) = \text{rect}\left(\frac{t}{T}\right) \cos\left[\frac{2\pi}{b} \ln\left(1 + \frac{b}{T_o} t\right)\right]$$

where b is the rate of the hyperbolic frequency modulation:

$$f(t) = \frac{1}{T_o + bt} \qquad -\frac{T}{2} \leq t \leq \frac{T}{2}$$

From eq. (3), the taps of the interdigital array must correspond to the times:

(21) $$t_n = \frac{T_o}{b}\left(e^{n\frac{b}{2}} - 1\right) \qquad n \text{ integer}, \ |t_n| \leq \frac{T}{2}$$

where the instantaneous phase of g(t) is multiple of π.

The number of taps in the interdigital array is given now by:

(22) $$N = \frac{2}{b} \ln\left(1 + b \frac{T}{T_o - b\frac{T}{2}}\right)$$

where T is the signal duration.

When a reduction of the factor k is allowed, the sampling times are given by:

(23) $$t_n^k = \frac{T_o}{b}\left(e^{kn\frac{b}{2}} - 1\right)$$

An hyperbolic group delay vs. frequency characteristic (fig.9) over the frequency range 7,75Mhz ÷ 10.25Mhz was achieved with an interdigital array with the electrodes spaced according to the delay (21).

The passive-generated signal and the matched-filter output are shown in fig. 10.

It can be easily verified that also in this case the number of taps (22) can be reduced of the factor k = 5.

The output signal obtained using an interdigital configuration with the minimum number of taps is shown in fig. 11.

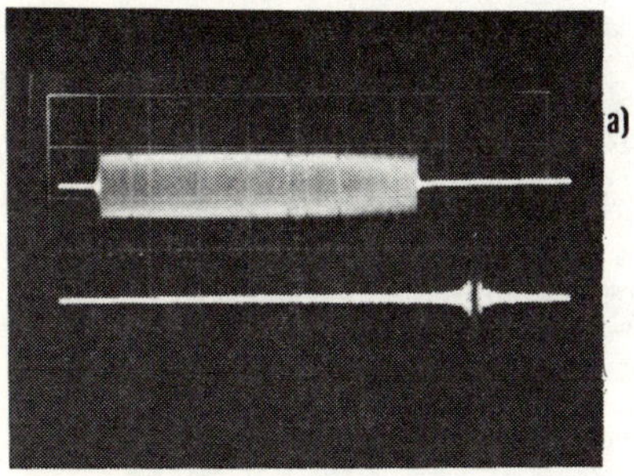

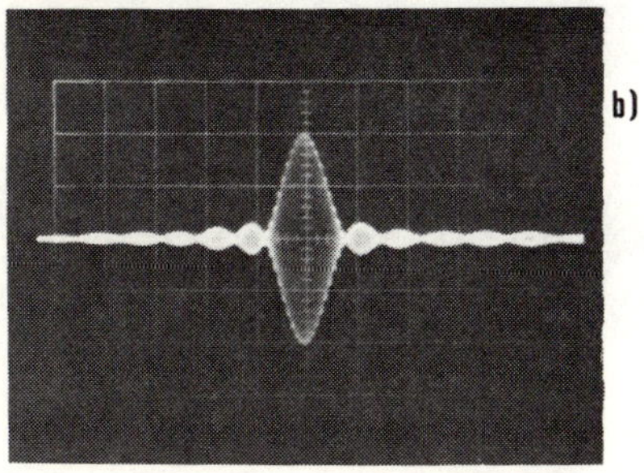

Fig. 10 - Acoustic surface-wave matched filter for hyperbolic frequency-modulated signal processing: k = 1.
a) Top: passive-generated pulse. Bottom: correlation pulse (5 μsec/div).
b) Details of the correlation pulse (0.5 μsec/div).

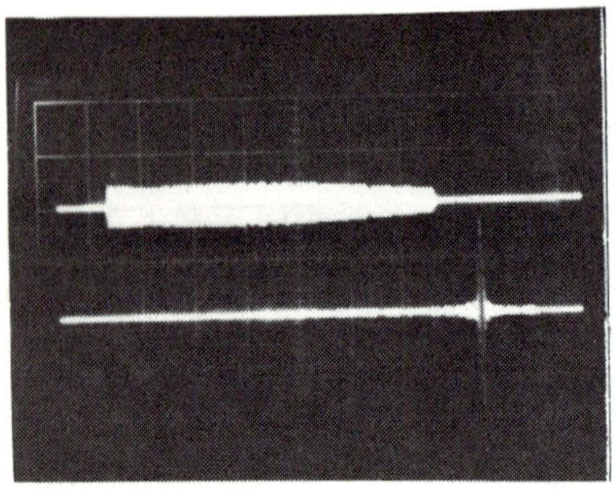

a)

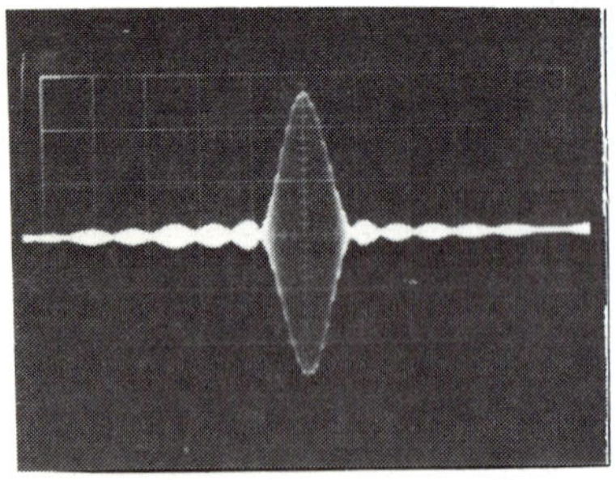

b)

Fig. 6 - Measured group delay vs. frequency chara cteristic of the acoustic surface-wave matched filter for linear frequency-modulated signal processing.

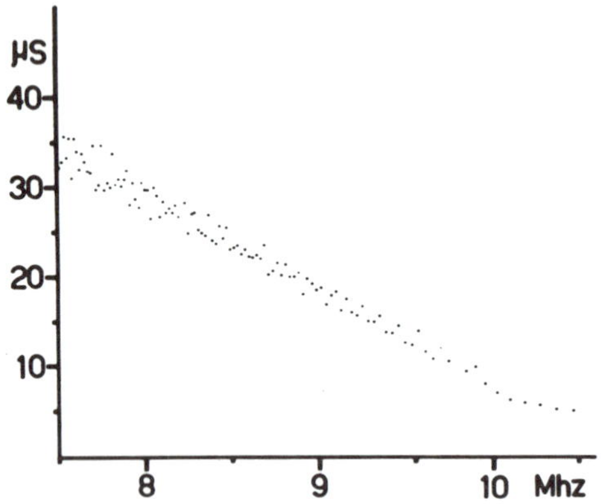

Fig. 7 - Acoustic surface-wave matched filter for linear frequency-modulated signal processing: k=1
a) Top: passive generated pulse
Bottom: correlation pulse (5 µsec/div)
b) Details of the correlation pulse (0.5 µsec/div).

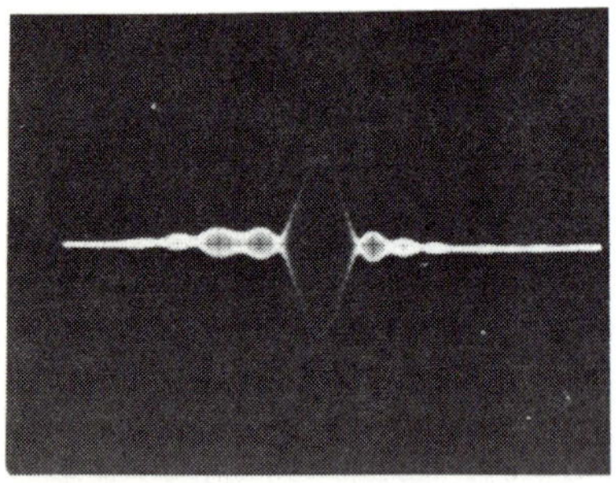

Fig. 8 - Acoustic surface--wave matched filter for linear frequency-modulated signal processing: k = 5. Details of the correlation pulse (0.5 µsec/div).

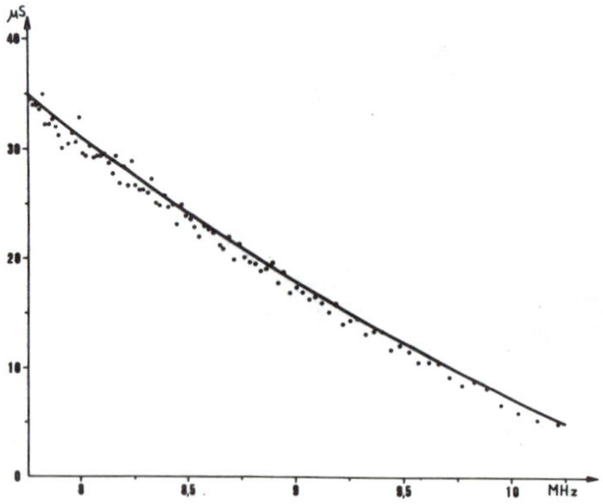

Fig. 9 - Measured group delay vs. frequency characteristic of the acoustic surface-wave matched filter for hyperbolic frequency-modulated signal processing.
The full line shows the expected delay.

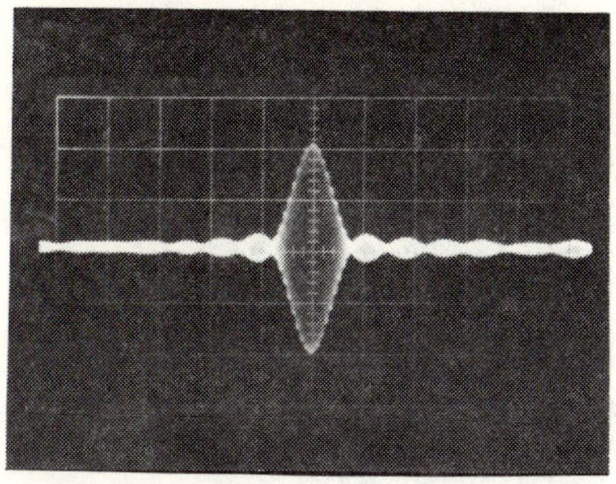

Fig. 11 - Acoustic surface-wave matched filter for hyperbolic frequency-modulated signal processing: k=5. Details of the correlation pulse (0.5 μ sec/div).

CONCLUSIONS.

- Planar processing techniques, familiar to people working in integrated circuit design, find a new promising application in the development of acoustic surface-wave transversal filters.
- It is the accessibility of surface waves to tapping that makes surface waves so attractive for this type of filtering.
- The filter response function is synthesized by a suitable configuration of tapping transducers, consisting of conductive electrodes deposited on a piezoelectric substrate. The fabrication technique simply requires the use of standard photolithographic procedures.

In this paper the sampling procedure which is the basis for the design of the tapping electrode configuration has been outlined, and some experimental results of acoustic surface wave matched filters have been reported.

REFERENCES

(1) W.D.SQUIRE, H.J.WHITEHOUSE, J.M.ALSUP: Linear signal processing and ultrasonic filters. - "IEEE Trans. on Microwave Theory and Techniques", vol. MTT-17, pp. 1020÷1040, 1969.

(2) C.ATZENI, L.MASOTTI: A new sampling procedure for the synthesis of linear transversal filters. - "IEEE Trans. on Aerospace and Electronic Systems", May 1971 (in print).

(3) LORD RAYLEIGH: On waves propagated along the plane surface of an elastic solid. - Proc. London Math. Soc., vol. 17, pp. 4÷11, 1885.

(4) R.M.WHITE, F.W.VOLTMER: Direct piezoelectric coupling to the surface elastic waves. - "Appl. Phys. Letters", vol. 7, pp. 314÷316, 1965.

(5) R.M.WHITE: Surface elastic waves. - Proc. IEEE", vol. 58, pp.1238÷1276, 1970.

(6) G.L.TURIN: An introduction to matched filters. - "IRE Trans. on Information Theory", vol. IT.6, pp. 311÷329, 1960.

(7) C.ATZENI, L.MASOTTI, E.TEODORI: Acoustic surface-wave matched filters. - "Alta Frequenza", N.5, vol. 40, May 1971 (in print).

(8) C.ATZENI: Sensor number minimization in acoustic surface-wave matched filters. - "IEEE Trans. on Sonics and Ultrasonics", vol. SU.18, October 1971 (in print).

(9) KROSZCZYNSKI, J.J.: Pulse compression by means of linear-period modulation. - "Proc. IEEE", vol. 57, pp. 1260÷1266, 1969.

SURFACE WAVE TRANSDUCER ARRAY DESIGN USING TRANSVERSAL FILTER CONCEPTS

Jeffrey M. Speiser and Harper J. Whitehouse
Naval Undersea Research & Development Center
San Diego, California 92132

INTRODUCTION

The transversal filter concept permits a unified approach to the design of line arrays of transducers.[1] Such line arrays play a central role in the implementation of signal-processing functions using elastic surface waves.

Surface elastic wave devices are under development at our laboratory for performing linear and bilinear signal-processing functions including serial access memory, matched filtering, spectral analysis, and crosscorrelation or crossconvolution.

In the past, the performance of such devices has been limited by the available transducers rather than the fundamental limitations imposed by the propagation medium. This paper shows that coded transducer arrays may be designed which circumvent the previously assumed tradeoffs between bandwidth and insertion loss, between bandwidth and directionality, and between time-bandwidth product and insertion loss.

THE TRANSVERSAL FILTER APPROXIMATION FOR TRANSPARENT ACOUSTIC ARRAYS

The general structure which we will consider will consist of one or more line arrays of transducer elements. The transducer elements will be treated as acoustically transparent, thus permitting the use of the principle of superposition. The transducer elements will be assumed to have the same electrical-to-acoustic or acoustic-to-electrical impulse response apart from a weighting in magnitude and sign. (The weighting may be accomplished either electrically or in the design of the transducer element itself.) These approximations lead us to regard the transducer array as a tapped delay line with lightly-coupled, nonreflecting taps, also known as a transversal filter or Kallmann filter (fig. 1).

The electrical response of the receive transducer array to an applied acoustic impulse is thus the summation of the individual element (tap) responses:

$$h_R(t) = \sum_i r_i u\,(t - \tau_i) \qquad \text{(apart from a pure time delay)}$$

The transfer function is thus

$$H_R(f) = U(f) \sum_i r_i e^{-j2\pi f \tau_i}$$

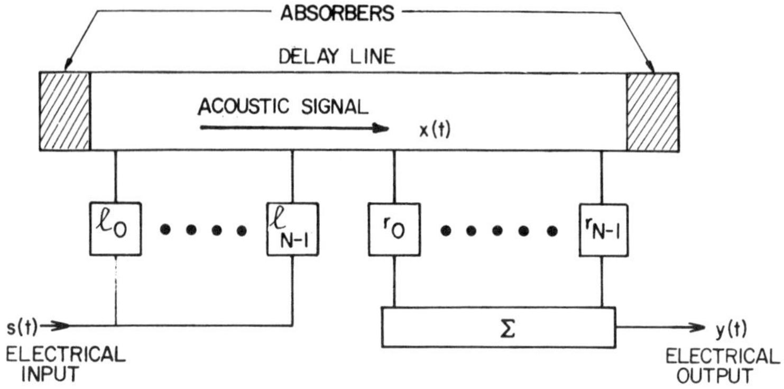

Figure 1. Cascaded Transversal Filter Equivalent of a Launch Transducer Array and a Receive Transducer Array on a Common Acoustic Path.

where

$$H_R(f) = \int h_R(t)\, e^{-j2\pi ft}\, dt$$

and

$$U(f) = \int u(t)\, e^{-j2\pi ft}\, dt$$

If the taps are equally spaced,

$$\tau_k = kd$$

and

$$H_R(f) = U(f) \sum_k r_k e^{-j2\pi fkd},$$

where U(f) is the transfer function of a single receive transducer element, and $\sum_k r_k e^{-j2\pi fkd}$ is the transfer function of a tapped delay line with ideal taps.

Note that if

$$r(t) = \Sigma r_k\, \delta(t - kd),$$

then

$$R(f) = \int r(t)\, e^{-j2\pi ft}\, dt = \sum r_k \int \delta\,(t - kd)\, e^{-j2\pi ft} dt = \sum r_k e^{-j2\pi fkd}.$$

That is, the receive transducer array may be thought of as an ideal transversal filter with impulse response $r(t) = \Sigma r_k\, \delta(t - kd)$, followed by a filter with impulse response u(t) characteristic of a single receive element.

Let v(t) = the acoustic response of a single launch transducer element to an electrical impulse. Then, the acoustic response of the entire launch transducer to an electrical impulse is (apart from a pure time delay)

$$h_L(t) = \sum_k \ell_k\, v(t+kd).$$

The corresponding transfer function is

$$H_L(f) = V(f) \sum_k \ell_k\, e^{j2\pi fkd}$$

This may be thought of as an ideal transversal filter with impulse response

$$\ell(t) = \sum_k \ell_k \, \delta(t+kd),$$

followed by a filter with impulse response, v(t), characteristic of a single launch element. The overall transfer function of the device is therefore

$$H(f) = H_L(f)\, H_R(f) = U(f)\, V(f) \sum_k \ell_k e^{j2\pi fkd} \sum_m r_m e^{-j2\pi fmd}$$

Define $W(f) = \sum_k \ell_k e^{j2\pi fkd} \sum_m r_m e^{-j2\pi fmd}$. Then, the overall transfer function is H(f) = U(f) V(f) W(f), where the factors U(f) and V(f) are determined solely by the choice of the launch and receive transducer elements, and the factor W(f) is determined solely by the choice of launch and receive weighting sequences.

In the time domain, the impulse response is h(t) = u(t) * v(t) * w(t), where * denotes convolution, and w(t) is the inverse Fourier transform of W(f).

$$W(f) = \sum_k \sum_m \ell_k r_m e^{j2\pi fd(k-m)} = \sum_s w_s e^{j2\pi fds}$$

where

$$w_s = \sum_{k-m=s}\sum \ell_k r_m = \sum_m \ell_{m+s} r_m .$$

That is, w_s is the crosscorrelation sequence of the launch and receive weight sequences. In the time domain

$$w(t) = \int W(f)\, e^{j2\pi ft}\, df = \sum_s w_s \, \delta(t+sd)$$

or

$$w(t) = \sum_s \left(\sum_m \ell_{m+s} r_m\right) \delta(t+sd) .$$

That is, modulo the impulse response of the launch and receive transducer elements, the impulse response of the overall structure is a pulse train with weight sequence given by the crosscorrelation sequence of the launch and receive weight sequences. This will be called the coded impulse response of the device.

THE FIELD-DELINEATED TRANSDUCER ELEMENT

The transducer elements used in the launch and receive arrays must be suitably chosen to insure the applicability of the above analysis.

If the electric field produced by a transducer element is entirely contained within the boundaries of that element, then the net electric field is a superposition of spatially separated terms, with each term corresponding to a single transducer element. This condition is satisfied by the "field-delineated" transducer element[2] shown in the dashed portion of fig. 2.

Figure 3 shows the acoustic response of the field-delineated transducer shown in figure 2. The dashed portion of figure 3 is the impulse response of the basic field-delineated element. Note that the impulse response indicated in figure 3 is that of a single element convolved with the time reversal of the weight sequence.

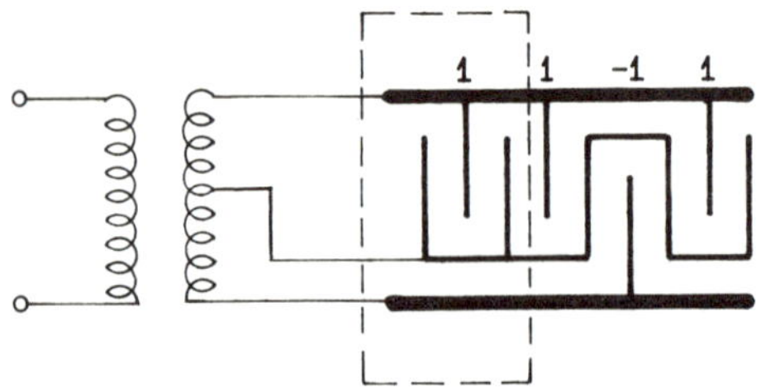

Fig. 2. Field-Delineated Transducer with Weight Sequence 1, 1, -1, 1.

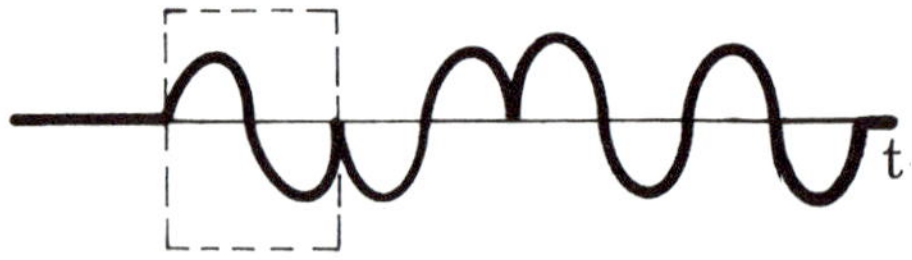

Fig. 3. Acoustic Response to an Electrical Impulse of the Field-Delineated Transducer of figure 2.

CODED TRANSDUCER ARRAYS

Two special choices for the launch and receive transducers are of particular interest. In the first, the launch transducer consists of a single element, and the coded impulse response is just the code on the receive transducer.

$$h(t) = u(t) * v(t) * \sum_m r_m \delta(t - md)$$

or

$$W(f) = U(f)\, V(f) \sum_m r_m e^{-j2\pi fmd}$$

In this case, a filter is designed by specifying its desired impulse response, $h_d(t)$, and choosing the receive transducer weights so as to be equally-spaced samples of h_d. Provided that the conditions of the sampling theorem are met, $h(t) = h_d(t)$.

In the second case of interest, the launch and receive transducers have the same weight sequences. In this case, the coded impulse response of the device is the autocorrelation function of the weight sequence.

$$h(t) = u(t) * v(t) * \sum_s \left(\sum_m r_m r_{m+s} \right) \delta(t+sd)$$

where r_m is a sequence of real numbers.

If we were trying to design a delay line with pulse-in/pulse-out response, the best we could do would be to make

$$\sum_m r_m r_{m+s} = \begin{Bmatrix} \Sigma r_m^2, & s=0 \\ 0, & s \neq 0 \end{Bmatrix},$$

in which case, $h(t)$ would be a scaled copy of $u(t) * v(t)$.

No sequence of more than one term has an autocorrelation function which is completely free of sidelobes, but several approximate solutions are known. The Barker codes[3] are binary sequences (sequences of + 1's and - 1's) whose autocorrelation function sidelobes do not exceed one in absolute value. No Barker codes are known with length exceeding 13. Huffman's "impulse equivalent" codes[4] eliminate all but the sidelobes occurring at maximum positive and negative shifts. Unfortunately, the longer Huffman codes use very nonuniform weights, and no Huffman code has been exhibited for which $\Sigma r_m^2/\max r_m^2$ is even as large as the value of 13 attainable with the longest known Barker code. The quantity $\Sigma r_m^2/\max r_m^2$ may be interpreted as the gain produced by using coded launch and receive transducers in a delay line, instead of a single launch element and a single receive element.

MULTIPLE-PATH FILTERS

By utilizing the additional degrees of freedom provided by using more than one acoustic path, we can eliminate the design constraints imposed in the previous section.

A multiple-path launch transducer combined with a multiple-path receive transducer is illustrated below (fig. 4).

The K launch transducers have their left ends terminated in acoustic absorbers, and the K receive transducers have their right ends similarly terminated.

If $H_{Lk}(f)$ denotes the transfer function of the kth launch transducer and $H_{Rk}(f)$ denotes the transfer function of the kth receive transducer, then the overall transfer function of the device (apart from the propagation delay) is $H(f) = \sum_k H_{Lk}(f)\, H_{Rk}(f)$. Let the kth launch transducer have weight sequence

$$\ell_{k,1} \quad \ell_{k,2} \quad \cdots \quad \ell_{k,N}$$

and let the receive transducer have weight sequence

$$r_{k,1} \quad r_{k,2} \quad \cdots \quad r_{k,N}.$$

Then, the transfer function for the kth path is

$$H_{Lk}(f)\, H_{Rk}(f) = U(f)\, V(f) \sum_s \left[\sum_m \ell_{k,m+s}\, r_{k,m} \right] e^{j2\pi fsd} .$$

The overall transfer function is

$$H(f) = U(f)\, V(f) \sum_s \left[\sum_k \sum_m \ell_{k,m+s}\, r_{k,m} \right] e^{j2\pi fsd} .$$

In order to have $H(f) = c\, U(f)\, V(f)$, we need

$$\sum_k \sum_m \ell_{k,m+s}\, r_{k,m} = \begin{cases} c, & s=0 \\ 0, & s\neq 0 \end{cases} ,$$

or

$$\sum_{k=1}^{K} L_k \otimes R_k = c\, \delta_{0s} ,$$

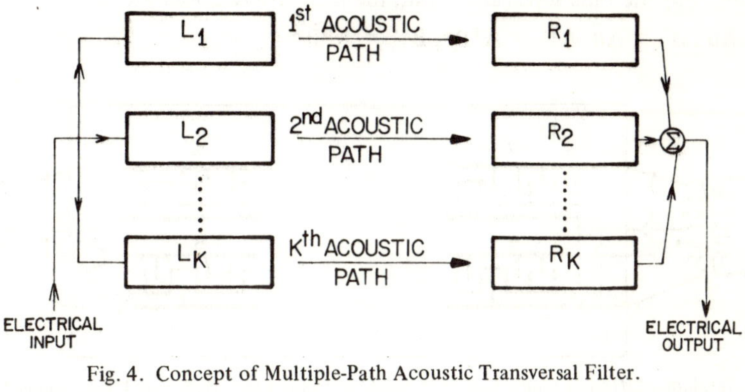

Fig. 4. Concept of Multiple-Path Acoustic Transversal Filter.

where

$\otimes$ denotes correlation
s is the shift index
L_k is the kth launch weight sequence
R_k is the kth receive weight sequence and
δ_{jk} is the Kronecker delta.

SURFACE WAVE TRANSDUCER WITH HIGH FRACTIONAL BANDWIDTH USING GOLAY SEQUENCES

We have just shown that in order for a multiple path filter to have pulse-in/pulse-out response, the sum of the weight crosscorrelation sequences for the several paths must be a delta function. The simplest nontrivial solution is provided by the "complementary series" of M. Golay.[5] A complementary series is a pair of binary sequences A, B, such that the sum of their autocorrelation functions is a delta function.

A simple Golay sequence pair is A = 1, 1
B = 1, -1

Golay sequences of arbitrarily great length may be generated using Golay's algorithm

$$A' = AB = a_0, a_1, \ldots a_{N-1}, b_0, b_1, \ldots . b_{N-1}$$

$$B' = A\overline{B} = a_0, a_1, \ldots a_{N-1}, -b_0, -b_1, \ldots -b_{N-1}$$

where the overbar denotes replacing each element in the sequence by its negative. If A and B are complementary sequences, then A′ and B′ will be complementary sequences of twice the length.

If in the multiple-path filter of fig. 4 we let $L_1 = R_1 = A$ and let $L_2 = R_2 = B$, where A and B are complementary sequences, the filter's impulse response is an approximation to a delta function. It is exactly 2N times the impulse response which would be obtained using a single launch element and a single receive element, where N is the length of the Golay sequence pair.

The Golay-coded delay line (fig. 5) uses two acoustic propagation paths to produce the equivalent of a line using a single launch element and a single receive element, but with reduced insertion loss.

Schweitzer[6] proposed the addition of a second pair of launch transducers and a second pair of receive transducers (fig. 6) to produce two independent delay lines, each using the same two acoustic paths.

In order that the two delay lines have pulse-in/pulse-out response and zero crosstalk, the weight sequences must satisfy:

$$A \otimes A + B \otimes B = c_1 \delta_{0s}$$

$$A^* \otimes A^* + B^* \otimes B^* = c_2 \delta_{0s}$$

$$A \otimes A^* + B \otimes B^* = 0$$

where c_1 and c_2 are constants.

A trivial solution is A=1, B=1, A*=1, B*=-1.

The present authors have shown[7] that given any solutions of the above equations, longer sequences satisfying the same equations can be generated by using Golay's time series algorithm or his interleaving algorithm. For the time series algorithm, the recursive relations are:

$$A' = AB\ , B' = A\overline{B}, A^{*\prime} = A^*B^*, B^{*\prime} = A^*\overline{B^*}$$

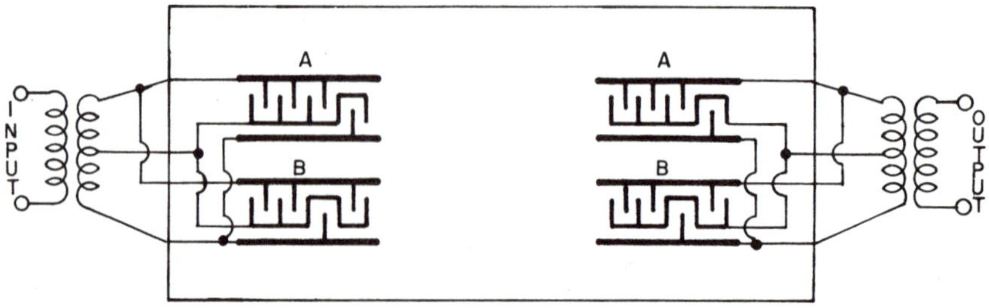

Fig. 5. Implementation of Golay-Coded Delay Line Using Field-Delineated Transducer Elements with A = 1, 1, 1, -1 and B = 1, 1, -1, 1.

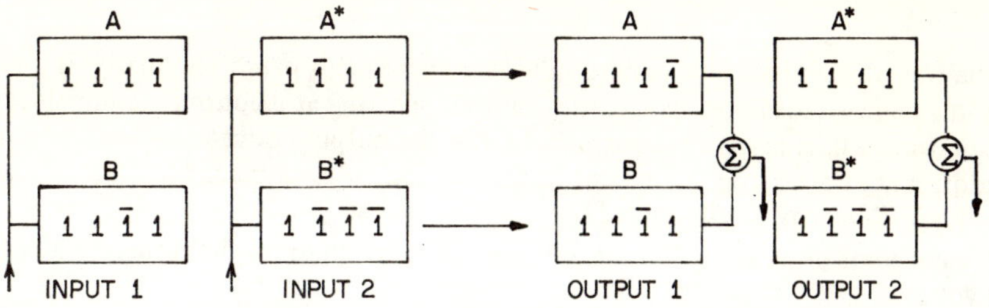

Fig. 6. Delay Line Structure Using Noninteracting Golay Sequences.

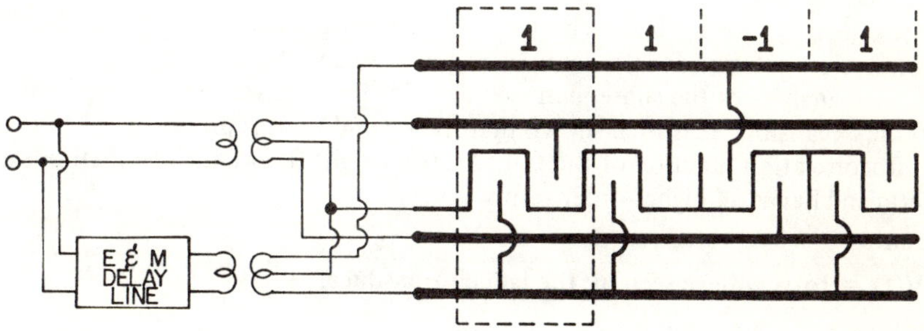

Fig. 7. Coded Unidirectional Launch Transducer Implementation with Weight Sequence 1, 1, –1, 1.

An alternative method of solving the above equations has been found by Tseng[8], and generates the same sequences.

Considerations of dimensionality[7] show that no more than two Golay sequence pairs occur in any mutually noninteracting set.

In general, if $K \geqslant 2$ acoustic paths are utilized, sets of K noninteracting sequences may be constructed[9] using recursive algorithms similar to the Golay algorithms. If K is the order of a Hadamard matrix, the sequences may be chosen to be binary, and hence are particularly suitable for multiplex delay line applications.

BROADBAND AND CODED UNIDIRECTIONAL TRANSDUCERS

A unidirectional, surface-wave transducer has been proposed for special delay line applications[10]. We can design coded unidirectional transducers by choosing as our basic transducer element a broadband, unidirectional version of the field-delineated transducer element, as illustrated in the dashed section of fig. 7.

Such a coded transducer may be thought of as a spatial replication of weighted, unidirectional transducer elements, and hence its impulse response is simply the convolution or correlation of the weight sequence with the response of the basic unidirectional transducer element.

The basic unidirectional element is a pair of contiguous, field-delineated transducer elements of opposite polarity, driven with a relative delay equal to the acoustic propagation time between them. The direction of propagation is determined by which member of the pair is delayed with respect to the other. This required auxiliary delay is small enough that it can be conveniently realized with an electromagnetic delay line. Only one such auxiliary delay line is needed for a coded unidirectional transducer, and the delay required does not increase with the number of elements in the weight sequence, so that the bandwidth is not necessarily reduced for long weight sequences.

The Golay-coded transducers of fig. 5 may be implemented with unidirectional transducer elements to produce a broadband delay line with unidirectional propagation.

BACK EMF AND HIGH COUPLING SUBSTRATES

If a launch transducer is driven by a finite impedance source, the driving voltage at the transducer terminals will be perturbed by the acoustic wave which the transducer launches. If the source impedance is high and the coupling is high, this perturbation voltage may result in a perturbed acoustic output, invalidating the transversal filter approximation used earlier in this paper. Similarly, if a receive transducer is terminated in a nonzero load impedance, the voltage appearing at its output terminals may launch a secondary acoustic wave which results in a perturbation of the output voltage.

In addition to the obvious expedient of driving with a low-impedance source, our analysis will show that the back EMF may be eliminated in two other ways:

a) Golay coded or other complimentary transducers driven in parallel generate identically cancelling perturbation voltages.

b) Identical transducers coupled through an ideal gyrator generate identically cancelling perturbation voltages. (This method requires a broadband gyrator with low insertion loss.)

In the following phenomenological analysis of the high-coupling interaction, it is assumed that the transduction is described by linear relations between the electrical and acoustic variables. This assumption leads to a Neumann series solution for the acoustic response to an electrical impulse.

The detailed analysis of the subsequent section uses the transversal filter approximation to generate the first term in the expansion. It is then shown that the second term is proportional to the positive-time portion of the autocorrelation function of the transducer's weight function. The methods a) and b) above eliminate this term and hence all higher-order terms.

Notation:

$V = V(t)$ = total voltage seen by the launch transducer

$V_{in} = V_{in}(t)$ = source voltage

$V_A = V_A(t)$ = acoustically-generated voltage

$A = A(x,t)$ = acoustic response (displacement, velocity, etc.)

L, M – linear, time-invariant operators

I = identity operator

Launch Transducer Analysis:

$$V = V_{in} + V_A$$

$$V_A = L\,A$$

$$A = M\,V$$

$$V = V_{in} + L\,M\,V \quad \Big\} \text{ a Fredholm integral equation}$$

Exact Solution:

$$V = (I - LM)^{-1}\,V_{in}$$

$$A = M\,V = M\,(I - LM)^{-1}\,V_{in}$$

Neumann Series Solution:

If the norm of LM is less than 1, then $(I - LM)^{-1} = I + LM + (LM)^2 + (LM)^3 + \ldots$

Moderate Coupling Approximation:

If the norm of LM is much smaller than one,

$$(I - LM)^{-1} \approx I + LM$$

A similar analysis applies to a receive transducer.

DETAILED ANALYSIS OF LAUNCH TRANSDUCER USING MODERATE COUPLING ASSUMPTION:

Normalize units so that propagation velocity = 1. Find acoustic response and electrical perturbation for an applied electrical impulse:

$$V_{in} = \delta(t)$$

$$A = M\,V \approx M\,V_{in} = K_1\,H(t)\,[W(x+t) + W(x-t)] \quad \text{first approximation to acoustic response}$$

$W(x)$ = transducer weight function

$H(t)$ = Heaviside step function

$$
\begin{aligned}
V_A &= L\,A = K_2 \int W(X)\,A(X,t)\,dX \\
&= K_1 K_2\,H(t) \int W(X)\,[W(X+t) + W(X-t)]\,dX \\
&= 2\,K_1\,K_2\,H(t)\,R_W(t)
\end{aligned}
$$

where

$$R_W(t) = \int W(U)\,W\,(U+t)\,dU = \text{the autocorrelation function of W}$$

that is,

$$V_A(t) \approx \begin{cases} 0, & t<0 \\ K\,R_W(t), & t>0 \end{cases}$$

This perturbation voltage has been experimentally observed for a periodic transducer using a Lithium Niobate substrate, and the back EMF cancellation for Golay-coded transducers has been demonstrated indirectly, by observing that the same Golay-coded transducers on Quartz and Lithium Niobate produce impulse responses having identical shapes[11].

A PROPOSED WIDEBAND CROSSCONVOLVER

Golay-coded transducers may be used for broadband transduction in bilinear as well as linear devices. A simplified theory of nonlinear interactions in high-coupling substrates indicates that such transducers may be used to build a broadband version of the Quate-Kino acoustic crossconvolver or crosscorrelator[12,13]. The simplified model predicts that when a pair of collinear but oppositely-directed surface-waves pass under a transducer on a high-coupling substrate (fig. 8a), the transducer's electrical output consists of the sum of three groups of terms: a pair of terms which are linear in the two inputs, a pair of terms which are quadratic in the two inputs, and a term which is bilinear in the two inputs. The quadratic terms may be eliminated by a differential configuration (fig. 8b) analogous to the well-known quarter-square multiplier. If the launch transducers have zero d.c. response, then the spatial average of the linear terms will be zero. Therefore, if the receive transducer consists of a layer of uniform metalization, the linear terms will vanish during the time that the acoustic waves are entirely under the receive transducer.

If the quadratic and linear terms are eliminated, the device may be treated as bilinear, and its output may be described as $x_1 * x_2 * h$, where x_1 and x_2 are the two inputs, and h is the bilinear impulse response characterizing the device. If the response to a pair of applied delta functions itself approximates a delta function, then the device will be a broadband convolver. That is, assume a relation of the form

$$E = pz + q\,z^2$$

where

$$z = z_1 + z_2$$

then

$$
\begin{aligned}
E &= p\,(z_1 + z_2) + q\,(z_1 + z_2)^2 \\
&= [p\,z_1 + p\,z_2] + \quad q\,z_1^{\,2} + q\,z_2^{\,2} \quad + [2\,q\,z_1\,z_2]\,.
\end{aligned}
$$

For the configuration shown in figure 8c, if the two electrical inputs are delta functions, then while the acoustic waves are totally under the receive transducers, the spatially-integrated product appearing at the output is the sum of the autocorrelation functions of A and B. If A and B are coded via Golay sequences, this will be a delta function.

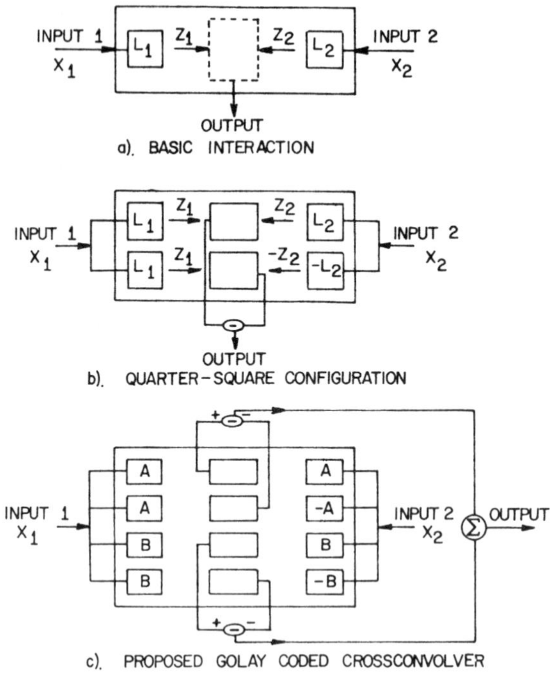

Fig. 8. Conceptual Evolution of Golay-Coded Crossconvolver.

REFERENCES AND FOOTNOTES

1. Squire, W. D., H. J. Whitehouse, and J. M. Alsup. Linear Signal Processing and Ultrasonic Transversal Filters. *IEEE Transactions on Microwave Theory and Techniques,* Volume MIT-17, Number 11, p. 1020-1040, Nov. 1969.
2. Whitehouse, H. J. A Wide-Band Low Insertion-Loss Multi-Electrode Surface-Wave Transducer, *1969 IEEE Ultrasonic Symposium.* Saint Louis, Mo., Sept. 24-26, 1969.
3. Barker, R. H. Group Synchronizing of Binary Digital Systems. *Communication Theory.* W. Jackson, Ed. London: Butterworth, 1953, p. 273-287.
4. Huffman, D. A. The Generation of Impulse-Equivalent Pulse Trains. *IRE Transactions on Information Theory,* Vol. IT-8, p. S10-S16, 1962.
5. Golay, M. J. E. Complementary Series. *IRE Transactions on Information Theory,* Vol. IT-7, p. 82-87, April 1961.
6. Schweitzer, B. P. Private communication.
7. Speiser, J. M. and H. J. Whitehouse. New Surface-Wave Transducer Codings, presented at the *1970 IEEE Ultrasonics Symposium,* San Francisco, Oct. 21-23, 1970.
8. Tseng, C. C. Signal Multiplexing in Surface-Wave Delay Lines Using Orthogonal Pairs of Golay's Complementary Sequences, *IEEE Transactions on Sonics and Ultrasonics,* Vol. SU-18, No. 2, p. 103-107, April 1971.
9. Whitehouse, H. J. Surface-Wave Transversal Filters, presented at the *IBM International Symposium on Surface Waves,* Yorktown Heights, N. Y. March 1970.
10. Collins, J. H. et al. Unidirectional Surface-Wave Transducer, *IEEE Proceedings,* Vol. 57, No. 5, p. 833-835, May 1969.
11. This has been experimentally verified by B. Schweitzer of Hughes Aircraft Co. and E. Stern of MIT Lincoln Laboratory.
12. Quate, C. F. and R. B. Thompson, Convolution and Correlation in Real Time with Nonlinear Acoustics, *Applied Physics Letters,* Vol. 16, No. 12, p. 494-496, 15 June 1970.
13. Luukkala, M. and G. S. Kino. Convolution and Time Inversion Using Parametric Interactions of Acoustic Surface Waves, *Applied Physics Letters,* Vol. 18, No. 2, p. 393-394, 1 May 1970.

INTERACTION OF LIGHT WITH ACOUSTIC SURFACE WAVES

A. Alippi, A. Palma, L. Palmieri, G. Socino
Istituto di Fisica, Università di Perugia, Italy
and
Istituto di Acustica "O. M. Corbino" C.N.R.
Via Cassia, 1216, 00189 Rome, Italy

1. Introduction

Interaction of light with elastic surface waves seems to stand as a simplified counterpart to bulk waves acousto-optical interaction in so far as the experimental task is concerned, but as a relatively more complex one when a thorough theoretical approach to it is attempted. Indeed, the thin layer of material where the energy carried by surface waves is confined greatly reduces the ensemble of problems related to geometrical misalignements and a simple exponential decay law of strains on depth takes place of complex radiation patterns proper to finite width acoustic beams. On the other hand, the inhomogeneous characteristics of Rayleigh waves toghether with the rippling of the propagation surface do both enhance difficulties present in solving the problem of light diffraction[1,2,3].

In the present work a theoretical approach to the problem is developed in the customary assumptions of neglecting speed and frequency of sound with respect to the ones of light and on assuming, too, that phase corrugation undergone by light wavefront is small, which condition - though being greatly restrictive in bulk wave analysis - is practically all times satisfied for surface waves.

Indeed, in crossing the free surface of a transparent solid substrate where an elastic surface wave is propagating, a light beam undergoes both a refraction effect at the discontinuity surface and a diffraction one which is jointly produced by the rippling of the surface and the local change of refractive index of the medium. Refraction has to be handled by referring to well- known Snell's laws and will give rise to deviation of the impinging light beam as well as to reduction of its light intensity. The two diffracting effects will be treated separately and will give rise, in the approximation of small phase corrugation, to an emerging wavefront split into three main portions, which correspond to two first order diffracted wavefronts and a zero undiffracted one.

According to what is outlined in Fig. 1 and coherently with the approximation above mentioned, we will only consider the first orders to be made up by light wavefronts which have been diffracted only once, either by internal periodic structure of strains (segment array) or by the grating corresponding to surface corrugation (sinusoidal line). The two wavefronts of each order have equal frequencies, for being undergone to the same amount of Doppler shift and different complex amplitudes respectively proportional to $V_{\pm 1}(\vartheta')$ and $J_{\pm 1}(\vartheta')$ which will be evaluated further on as function of the angle of incidence ϑ' of the light beam with the normal to propagation surface.

The light intensity of first diffraction orders, which is object of experimental processing and evaluation as well, is easily obtained by squaring the resulting field, which sums up the two complex amplitudes

$$I_1(\vartheta') = (E_o T_1)^2 \left| V_1(\vartheta') + J_1(\vartheta') \right|^2 \qquad (1)$$

where E_o is the amplitude of the impinging light and T_1^2 is the optical transmittance of the specimen surface for the angular geometry of first diffraction order.

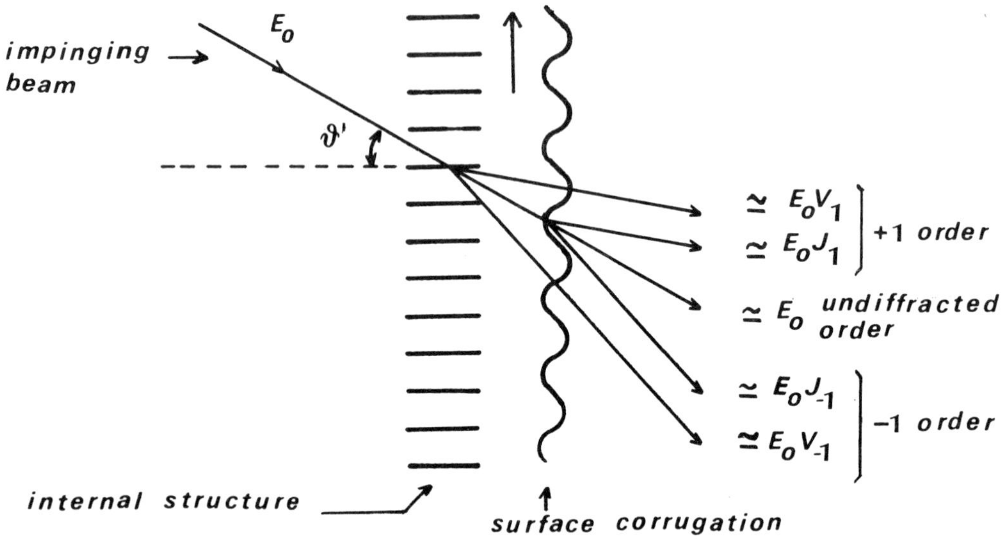

Fig. 1 - Simplified model of the two-steps interaction.

2. Theoretical approach.

For quantitative evaluation of the interaction problem we refer to the geometry sketched in Fig. 2. An infinitely extended plane wave of linearly polarized light with wave number k' and frequency ω, represented by the equation

$$E(x_1, x_3, t) = E_o\,(k' \sin\vartheta'\, x_1 + k' \cos\vartheta'\, x_3 - \omega t) \qquad (2)$$

impinges at angle ϑ' from inside on the plane free surface of a transparent solid medium where an elastic surface wave is travelling in x_1 direction with wave number K and frequency Ω. The particle displacement vector has components u_k which, in complex notation, are given by[4]

$$u_k(x_1, x_3, t) = A \exp i\,(K x_1 - \Omega t) \sum_1^3 j\, u_k^j \exp(K q_j x_3) \qquad (x_3 \leq 0,\ k = 1,2,3) \qquad (3)$$

where A is a real arbitrary constant, u_k^j are complex amplitudes and q_j dimentionless decay constants.

The surface wave locally produces a modification of the optical properties of the medium conveniently described in terms of variations ΔB_{ij} of the dielectric impermeability tensor B_{ij}^o, which are bound to local strain tensor $S_{kl} = 1/2\,(\partial u_k/\partial x_l + \partial u_l/\partial x_k)$ and photoelastic coefficients p_{ijkl} through equation

$$B_{ij} = p'_{ijkl} S_{kl} \qquad (i,j,k,l=1,2,3), \qquad (4)$$

where the prime stands for quantities to be evaluated in laboratory coordinate system. For evaluating the elasto-optical interaction, the elliptical central section of the modified indicatrix

$$(B_{ij}^o + \Delta B_{ij})\, x_i x_j = 1 \qquad (5)$$

has to be found normal to light propagation direction. Since the interaction geometry is so chosen that the $\bar{k}'$ vector lies on plane normal to x_2 axis, the ellipse is represented in the intersection plane, referred to orthogonal coordinates y_1, y_2 ($y_2 \equiv x_2$), by

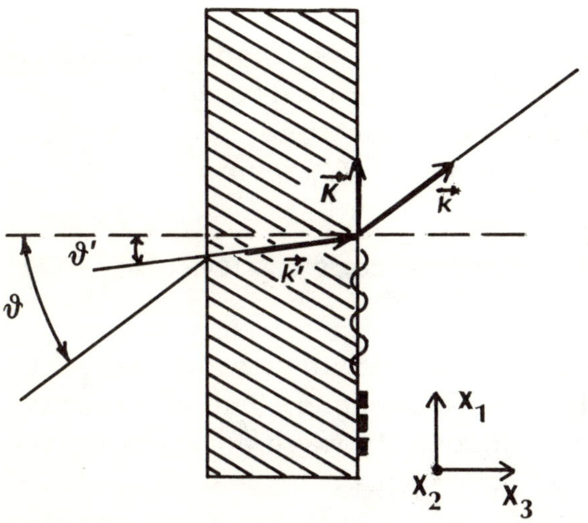

Fig.2 - Geometry of the interaction.

$$(B_{11}\cos^2\vartheta' + B_{33}\sin^2\vartheta' + B_{13}\sin 2\vartheta')\, y_1^2 + B_{22}\, y_2^2 + 2\,(B_{12}\cos\vartheta' + B_{23}\sin\vartheta')\, y_1 y_2 = 1 \qquad (6)$$

where $B_{ij} = B_{ij}^o + \Delta B_{ij}$. Directions of polarization for ordinary and extraordinary rays coincide with directions of the ellipse axes, whose lenghts are twice the corresponding refractive indices.

Inspection of Eqs. (4) and (6) suggests that surface waves generally cause the principal axes of the ellipse to change size and orientation, though rotation is only sensible in a restricted number of cases. In elastically isotropic materials, where elastic surface waves lack of sagittal component, the axes of the strain induced ellipse are parallel to y_1 and y_2, so that diffracted beams mantain the polarization of the incident light, if this is initially polarized along either one of the axes. For elastically anisotropic materials, too, strain induced rotation is not really sensible, unless for cases where the underformed ellipse degenerates in a circle and terms B_{12}, B_{23} differ from zero, too.

We actually neglect this last case and confine the analysis to cases of polarized impinging light with electric vector directed along either one of the ellipse axes. Any other polarization is easily reducible to a combination of these two cases. In what it follows, quantities which depend on polarization will carry a p index to signify that their value should be taken relative to that polarization, that will be eventually specified as p1 and p2 for polarization respectively parallel to axes y_1 and y_2.

After elasto-optical interaction has taken place in the medium, the main part of light intensity will be found at exit plane $x_3 = 0$ on a plane wavefront, that will constitute the undiffracted beam. A small fraction of impinging light will be however deflected from its original direction into plane waves which - coherently with the quantum nature of scattering processes - will have wave vectors $\bar{k}' + m\,\bar{K}$, with integer m, frequencies $\omega + m\,\Omega$ and complex amplitudes $V_m(\vartheta')$, what simply extends the notation already used for first diffraction orders alone. Though our final interest will be confined to determination of $V_1(\vartheta')$, a trial solution is given for the electric field E' in the medium vs. x_3 and ϑ' variables, which is made up of all V_m terms[5]

$$E'^p = E_o \sum_{-\infty}^{+\infty} {}_m\, V^p_m(x_3, \vartheta') \exp i\left[(k'\sin\vartheta' + mK)\,x_1 + k'\cos\vartheta'\,x_3 - (\omega + m\Omega)t\right] \quad (7)$$

where amplitudes V^p_m vary along x_3 as more photons scatter away from their original direction through interaction with surface phonons. The plane waves solution of Eq. (7) has to be substituted into Maxwell's wave equation for the polarized electric field in matter

$$c^2\,\nabla^2 E'^p = (n^p_o)^2\ \partial^2/\partial t^2\ (1 + 2\,\delta n^p/n^p_o)\ E'^p \quad (8)$$

where n^p_o is the refraction index of the unperturbed medium, c the velocity of light in vacuum and δn^p are just the variations of semiaxes of ellipse of Eq. (6), which,in case x_2 coincide with direction of one of the principle axes, reduce to

$$\delta n^{p1} = -(n^{p1}_o)^3\ (\Delta B_{11}\cos^2\vartheta' + \Delta B_{33}\sin^2\vartheta' + 2\,B_{13}\sin 2\vartheta')\,/\,2 \quad (9)$$

and

$$\delta n^{p2} = -(n^{p2}_o)^3\ \Delta B_{22}\,/\,2 \quad (9')$$

The following set of coupled equations is then obtained for quantities $V^p_m(x_3, \vartheta)$ in the approximation of neglecting term $\partial^2 V^p_m/\partial x_3^2$, for being V^p_m a slowly varying function on x_3, and terms Ω/ω and v_R/c with respect to unity, where v_R is the Rayleigh waves velocity,

$$\partial V^p_m/\partial x_3 + i\,\beta_m\,V^p_m + \xi^p(x_3, \vartheta')\,(V^p_{m+1} - V^p_{m-1}) = 0 \quad (10)$$

where

$$\beta_m = -\,K\,m\ (\sin\vartheta' + K m/2k)\,/\,\cos\vartheta' \quad (11)$$

and

$$\xi^p(x_3, \vartheta') = k\ \delta n^p(x_3)\,/\,2\,n^p_o\ \cos\vartheta' \quad (11')$$

where $\delta n^p (x_3)$ has to be taken as the amplitude of δn^p variation.

Solutions for Eq. (10) have been given by many authors for case of δn^p constant on a finite interval of x_3 axis and vanishing elsewhere.

In case of surface waves, parameter ξ^p presents an exponential decay along x_3 coordinate, so that integral $\int \xi^p \, d x_3$ can be easily calculated to be very small with respect to unity, what subsequently leads to $V_{\pm(m+1)} \ll V_{\pm m}$[6]. On limiting to cases m = ±1, corresponding to first diffraction orders, terms $V_{\pm 2}$ can be dropped in the corresponding equations which get then solved by

$$V^p_{\pm 1} (x_3, \vartheta') = \mp \exp(-i \beta_{\pm 1} x_3) \int_{-\infty}^{x_3} \xi^p (x'_3, \vartheta') \, V^p_o (x'_3, \vartheta') \exp(i \beta_{\pm 1} x'_3) \, dx'_3 \qquad (12)$$

Since the hypothesis was done that only a very small portion of the incident light is removed from its original direction, V^p_o can be assumed to be constant throughout the integration path and equal to one. The effect of the internal interaction is so described by expression

$$V^p_{\pm 1} (0, \vartheta') = \mp \int_{-\infty}^{0} \xi^p (x'_3, \vartheta') \exp(i \beta_{\pm 1} x'_3) \, dx'_3 \qquad (13)$$

which is the explicit form for the first term in second member of Eq. (1).

For what concerns the diffraction effect produced on the previously undiffracted beam of amplitude $V^p_o (0, \vartheta') \simeq 1$ by the rippling of the free surface due to transversal component $u_3 (x_1, 0, t)$ of the elastic wave, a trial solution for the electric field is given in the form

$$E^p = E_o \sum_{m=-\infty}^{+\infty} Z^p_m \exp i \left\{ k \left[\sin \vartheta_m x_1 + \cos \vartheta_m x_3 \right] - (\omega + m\Omega) t \right\} \qquad (14)$$

where $\vartheta_m = \sin^{-1} (n^p_o \sin\vartheta' + m K/k)$. The continuity conditions on the rippled surface between the electric fields of the impinging wavefront and the diffracted one of Eq. (14), determine coefficients Z^p_m, which result into

$$Z^p_m = T^p_m \; J_m \left[A \sum_{j=1}^{3} u^j_3 \, k \, (n^p_o \cos\vartheta' - \cos \vartheta_m) \right] \qquad (15)$$

where T^p_m are optical transmission coefficients of the specimen surface depending on angles ϑ' and ϑ_m and J_m is the m-th order Bessel function.

Expressions for $Z^p_{\pm 1}$ are so given by

$$Z^p_{\pm 1} = T^p_{\pm 1} \; J_{\pm 1} \left[A \sum_{j=1}^{3} u^j_3 \, k \, (n^p_o \cos\vartheta' - \cos \vartheta_{\pm 1}) \right] \qquad (16)$$

Following Eq. (1), the light intensity scattered into first orders due to contribution both of interface rippling and subsurface strain results to be

$$I_{\pm 1} = (E_o T^p_{\pm 1})^2 \left| \int_{-\infty}^{0} \xi^p \exp(i \beta_{\pm 1} x'_3) \, dx'_3 + J_{\pm 1} \left[A \sum_{j=1}^{3} u^j_3 \, k \, (n^p_o \cos\vartheta' - \cos\vartheta_{\pm 1}) \right] \right|^2 \qquad (17)$$

Inspection of Eq. (17) shows that the relative phase of the two contributing terms at second member does generally differ from 0 or π, unless the angle of incidence ϑ' coincide with the Bragg angle $\Theta = -\sin^{-1}(K/2k)$. The two contributing terms will add at Bragg angle constructively or destructively depending on photoelastic characteristics of material and strain structure of surface waves.

3. Experimental

Experimental determinations of first order light intensity I_1 have been performed vs. angle of incidence ϑ' for isotropic fused quartz substrate, X-cut α-quartz with Y propagation direction and Y-cut α-quartz with X propagation direction substrates.

Pulsed elastic waves were generated in the range 10-50 MHz by means of interdigital array transducers on piezoelectric α-quartz substrates and by hybrid transducers on fused quartz which was acoustically coupled with a lithium niobate crystal[7]. In all cases, polarization of light was kept normal to propagation direction of surface waves and measurements of light intensity were performed through photomultiplier detection and box-car integrator processing.

In elastically isotropic materials, for the selected polarization p2, which is mantained in diffracted light, parameter ξ^{p2}, which is responsible for the elasto-optical interaction, reduces to

$$\xi^{p2}(x_3, \vartheta') = -k\, n_o^3\, p_{12}\left[S_1(x_3) + S_3(x_3)\right] / 4\cos\vartheta' \qquad (18)$$

as obtained through Eqs. (4), (9'), (11') and use of conventional notation for contraction of indices. It is interesting to note that the sum of strains $S_1 + S_3$, representing the dilatation of the medium simply reduces to a single exponential decay term for an isotropic material

$$S_1(x_3) + S_3(x_3) = -A\,K\,(1 - q_1^2)\exp(K q_1 x_3) \qquad (19)$$

where $q_1 = \left[1 - (v_R/v_L)^2\right]^{1/2} < 1$, being v_L the velocity of longitudinal waves in fused silica. For usual values of surface waves power, argument of Bessel function in Eq. (17) is very small with respect to unity, so that J_1 can be approximated with first term of its series expansion. On considering positive first order alone, Eq. (17) then becomes

$$I_1 = (E_o A T_1^p k/2)^2 \left| K n_o^3 p_{12}(1-q_1^2)/2\cos\vartheta'(Kq_1 - i\beta_1) + q_1\left[1 - 2/(1+q_2^2)\right](n_o\cos\vartheta' - \cos\vartheta_1)\right|^2 \qquad (20)$$

where $q_2 = \left[1 - (v_R/v_S)^2\right]^{1/2} < 1$, being v_S the velocity of shear waves. It is easy to see from Eq. (20) that the relative phase $\varphi(\vartheta')$ with which the first contribution adds to the second is simply given by $\varphi = \mathrm{tg}^{-1}(\beta_1/K q_1)$.

In the range of acoustic frequencies that have been experimented, appro ximations hold that $\vartheta_1 \simeq \vartheta = \sin^{-1}(n_o \sin\vartheta')$ and $\beta_1 \simeq -K tg\vartheta'$, so that Eq.(20) can be thought as depending only on incident angle ϑ'. Using numerical data reported in the literature for elastic and photoelastic constants of fused silica[8,9], the theoretical curve of Eq. (20) has been calculated vs. internal angle ϑ'. In Fig. 3, for experimental convenience, the curve of Eq. (20) is plotted vs. external angle ϑ, corrected by transmittance factor $T^2(\vartheta)$ which takes account of refraction at first surface of the specimen. The experimental data relative to measurements performed at 20.0 MHz are also reported.

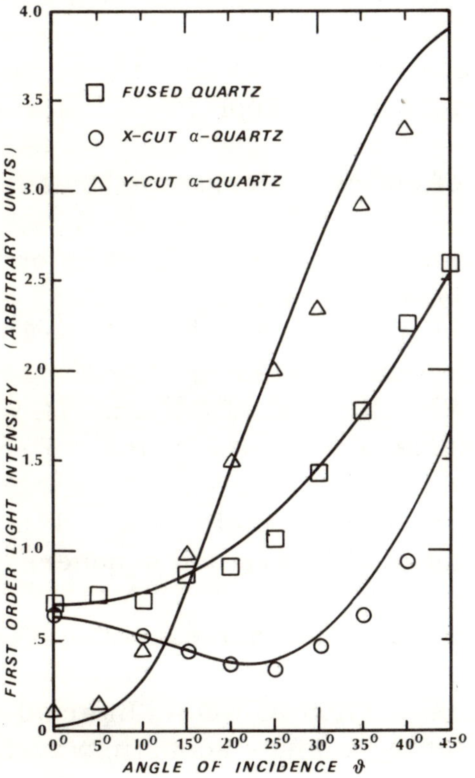

Fig. 3 - Angular dependence of first order light intensity for different substrates.

In both cases of anisotropic substrates considered, though sagittal component of surface wave is also present, strain induced rotation of the ellipse axes of Eq. (6) is still negligible. Therefore, if light polarization is so chosen as to be parallel to optical axis of the crystals, ξp^2 term will depend on ΔB_2 alone, which reduces to

$$\Delta B_2 = p_{13}(S_1 + S_3) \qquad (21)$$

for the crystal symmetry class 32 of α-quartz. Therefore, ξp^2 will formally have an expression identical to the one in Eq. (18), provided to substitute p_{12} with p_{13} and n_o with the refraction index n_e of extraordinary ray in α-quartz. In Fig. 3 the theoretical curve, evaluated through use of numerical data for amplitudes u_k^j given in ref. 4 and photoelastic constants of ref. 10, are reported together with the experimental data for X and Y-cuts at acoustic frequencies respectively equal to 18.8 MHz and 20.0 MHz.

It is worthwhile to note that the internal contribution to light diffraction adds destructively with the surface rippling contribution in all three cases considered. Indeed, in Y-cut case there is an almost complete destructive adding of the two contributions at normal incidence, which doesn't hold at increasing angles. In X-cut case a well defined minimum of light diffraction is present at a given angle of incidence, which corresponds to a maximum of the internal acousto-optical interaction produced by the π-phase change in dilatation S_1+S_3 occurring just below the surface for this particular crystal cut[11].

The single exponential law of decay on depth of dilatation S_1+S_3 proper to isotropic case causes the interaction to be dependent upon the q_1 and q_2 parameters alone, so making the specific case considered of general interest for any isotropic material.

References.

1) E. P. IPPEN - "Diffraction of light by acoustic surface waves", Ph. D. dissertation, University of California, Berkeley, 1968.
2) K. DRANSFELD and E. SALZMANN - "Excitation, detection, and attenuation of high-frequency elastic surface waves", Physical Acoustics, Academic Press, N. Y., 1970, vol. VII, p. 242.
3) E. G. H. LEAN and C. G. POWELL - "Optical probing of surface acoustic waves", Proceedings of the IEEE, 58, 1939 (1970).
4) K. A. INGEBRIGTSEN and A. TONNING - "Numerical data for acoustic surface waves in α-quartz and cadmium sulfide", Appl. Phys. Letters, 9, 16 (1966).
5) M. BORN and E. WOLF - "Principles of optics", Pergamon Press, N. Y., 1959, p. 593.
6) M. G. COHEN and E. I. GORDON - "Acoustic beam probing using optical techniques", Bell System Tech. J., 44, 693 (1965).
7) A. P. Van den HEUVEL, D. B. OWEN and S. G. JOSHI - "The hybrid transducer", IEEE Ultrasonics Symp., S. Louis, Missouri (1969).
8) F. W. VOLTMER, E. P. IPPEN, R. M. WHITE, T. C. LIM and G. W. FARNELL - "Measured and calculated surface-wave velocities", Proc. IEEE, 56, 1634 (1968).
9) R. W. DIXON - "Photoelastic properties of selected materials and their relevance for applications ot acoustic light modulators and scanners", J. Appl. Phys., 38, 5149 (1967).
10) T. S. NARASIMHAMURTY - "Photoelastic constants of α-quartz", J. Opt. Soc. Am., 59, 682 (1969).
11) A. ALIPPI, A. PALMA, L. PALMIERI and G. SOCINO - "Internal and surface contribution to light diffraction by surface acoustic waves", Appl. Phys. Letters, (1971), to be published.

ACOUSTIC SURFACE WAVE INVESTIGATIONS USING LASER LIGHT DEFLECTION

A. J. Slobodnik, Jr.
Air Force Cambridge Research Laboratories
Air Force Systems Command
L. G. Hanscom Field, Bedford, Massachusetts 01730

INTRODUCTION

The field of acoustic surface waves has received tremendous recent interest[1]. Many applications have been suggested particularly in the area of long-time-delay, wide-band-width signal processing devices. The subject of this paper is the detailed description of a laser probe[2,3] developed for studying microwave acoustic surface wave phenomena. This probe is essential for obtaining various design data for all surface wave devices operating in the microwave frequency region (where small wavelengths make the use of mechanical probes quite difficult).[4]

The basic operation of this device is illustrated schematically in Figure 1. Coherent light from a low power laser is directed onto a surface wave delay line. Due to the presence of the surface wave on the crystal acting as a diffraction grating, light is deflected into side lobes in addition to the specular direction. The intensities of these side lobes, which are directly related to the intensity of the surface wave, can then be monitored by a photomultiplier tube. This provides a convenient method of probing the behavior of the surface wave[5-7]. For example, by scanning the laser along the acoustic beam (i.e. on a path between the two transducers) one can measure the acoustic propagation loss on various materials. At microwave frequencies this is, of course, an important quantity as it is a significant portion of the overall insertion loss of any device. By investigating various materials with the laser probe, one can choose material having the lowest loss for a given device application. It has been found[2] that $LiNbO_3$ is a significantly lower loss material than quartz[8]. In addition, by scanning the laser across the acoustic beam the surface wave profile or shape of the beam can be determined[9]. This information also provides valuable design data as will be seen below.

The remainder of this paper will provide a detailed description of the laser deflection technique as well as give an introduction to some of its uses. Included will be sections devoted to acoustic propagation loss[2], nonlinear effects[10] (which provide particularly interesting device applications), beam steering effects[9] and a final section on the recently discovered[11] high frequency, high coupling, low beam steering cut of $LiNbO_3$.

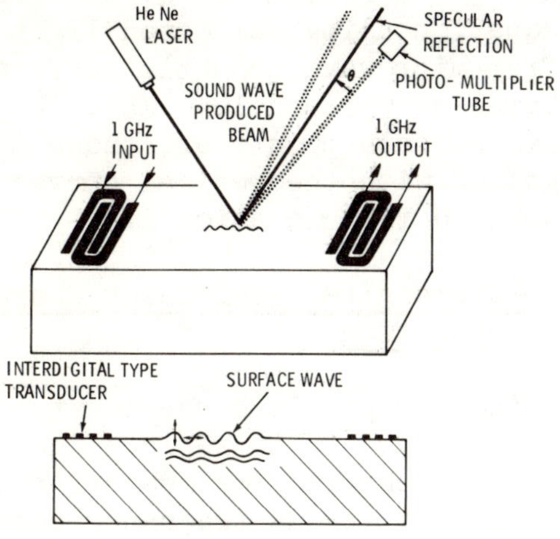

Figure 1 - Schematic diagram of a surface wave delay line and the laser probe used for investigating surface wave devices.

The Laser Probe

It is well known in electromagnetic theory that a periodically modulated surface causes plane waves of electromagnetic energy to be deflected to sidelobes in addition to the specular direction[12,13]. The angular directions of these sidelobes are given by the grating equation

$$\sin \theta_m = \sin \theta_o + \frac{m\lambda}{\Lambda_s} \qquad (1)$$

where λ is the wavelength of the incident energy and Λ_s is the surface wavelength. It can also be shown that the intensity of the deflected light is directly related to the intensity of the surface wave[12]. Thus using a low-power HeNe laser (λ = 6328A) as the incident electromagnetic energy, and placing a light detector at an angle corresponding to one of the sidelobes, yields a direct method of monitoring surface wave disturbances.

A detailed block diagram of the experimental laser probe system is shown in Figure 2. By square-wave

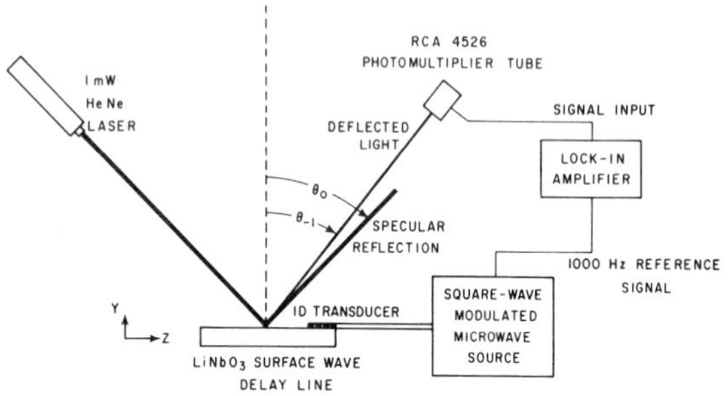

Figure 2 - Detailed block diagram of the laser probe system used for investigating microwave acoustic surface waves. The angles shown in this figure are defined in Equation1.

modulating the electromagnetic input to the interdigital transducer, the light deflected by the surface wave is also in the form of a square wave. This allows phase locked detection to be used and results in an improved signal to noise ratio. Using this system, measurements were easily made of acoustic surface wave power densities as low as 0.08 mw/mm at frequencies as high as 4 GHz.

The validity of Equation 1 was experimentally confirmed by measuring the angle of deflection of the first side lobe for light incident at various angles on surface waves propagating along the Z axis of Y-cut $LiNbO_3$. Experimental results for two frequencies (635 MHz and 1950 MHz) are compared with the theory in Figure 3. Agreement is excellent.

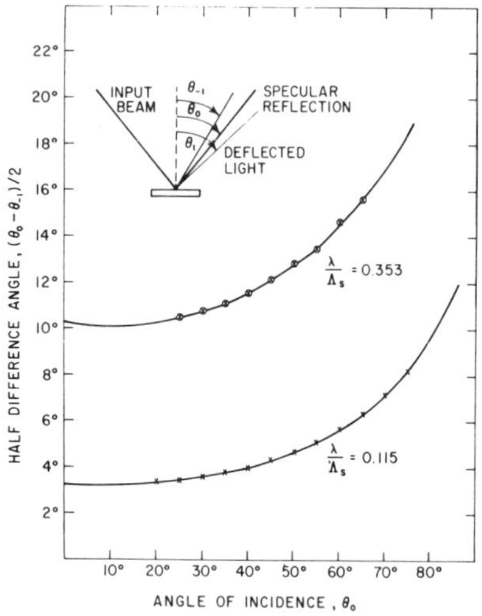

Figure 3 - Confirmation of the grating equation using deflected light from microwave acoustic surface waves at 635 and 1950 MHz. Experimental data points are shown together with theoretical curves derived from Equation 1.

Interdigital Transducer Charactertication

Another possible use for this optical method is to replace the standard technique for measuring insertion loss versus frequency characteristics of interdigital transducers[3]. Instead of a pair of transducers separated by a reasonable length of crystalline substrate, only one transducer and very little material are required. Alternatively, if one chooses to study standard delay lines, variations between the two transducers can be detected. Experimentally, both methods yield the same results versus frequency for the case illustrated in Figure 4. Relative laser measurements

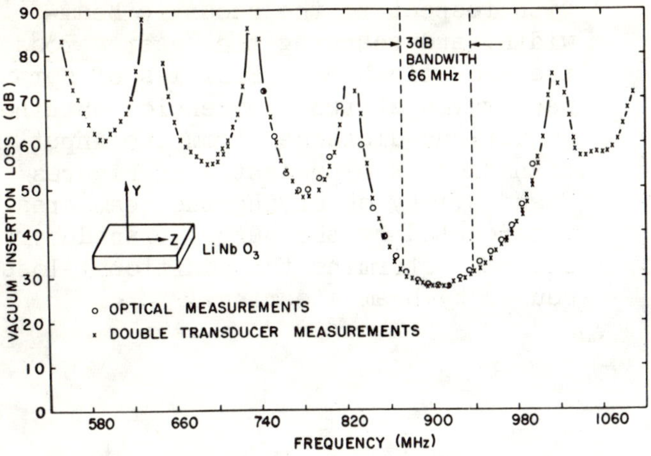

Figure 4 - Insertion loss versus frequency characteristics of a $LiNbO_3$ surface wave delay line measured by both double transducer and optical methods.

were made on both transducers of the delay line, and their sum (in decibels), normalized at 905 MHz to the transducer measurements, agrees with the double transducer technique over the entire frequency range of the data.

Attenuation and Beam Steering Measurements

In addition to RF conduction loss in the transducer fingers, two mechanisms frequently mentioned as contributing to total delay line insertion loss are acoustic attenuation[2] and beam steering[14,9]. In fact at microwave frequencies these are often the major contributions to insertion loss. The laser detection scheme allows convenient measurement of both these terms. To measure propagation losses one uses a laser beam wider than the acoustic beam (for averaging purposes) and then scans the light along the acoustic path. To avoid nonlinear effects[10] low input powers must be used. Illustrations of a low power attenuation curve as well as high power nonlinear effects are shown in Figure 5. The results of a collection of similar low power acoustic attenuation measurements are plotted in Figure 6 as a function frequency. Temperature dependence measurements have indicated that these results are very nearly down to the theoretical lower limit of thermal phonon scattering[15]. That is, loss contributions from crystalline defects or scratches and pits in the optical polish which are temperature independent, have been successfully eliminated. This is highly significant for practical device applications.

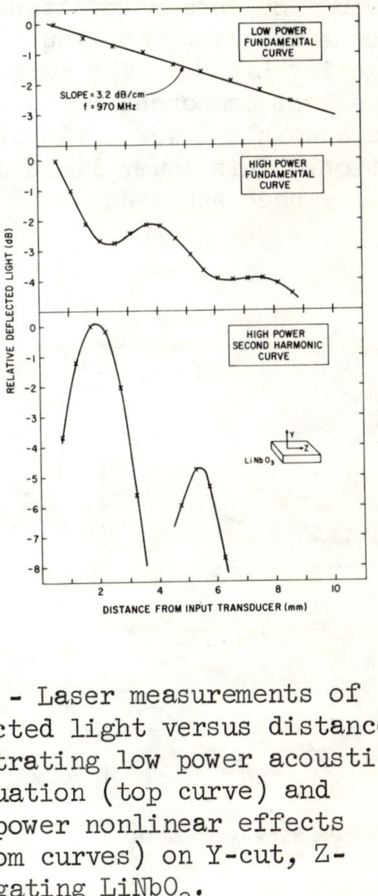

Figure 5 - Laser measurements of deflected light versus distance illustrating low power acoustic attenuation (top curve) and high power nonlinear effects (bottom curves) on Y-cut, Z-propagating $LiNbO_3$.

Beam steering arises from the misalignment of transducers from the desired pure mode crystalline axes causing the group and phase velocity vectors of the surface wave to lie in different directions[16]. The acoustic

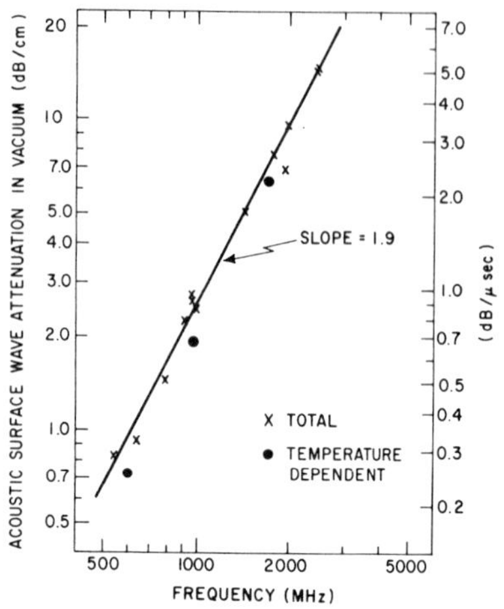

Figure 6 - Surface wave attenuation versus frequency on Y-cut, Z-propagating $LiNbO_3$. The temperature dependent component represents the absolute lower limit of attenuation. This lower limit has very nearly been achieved.

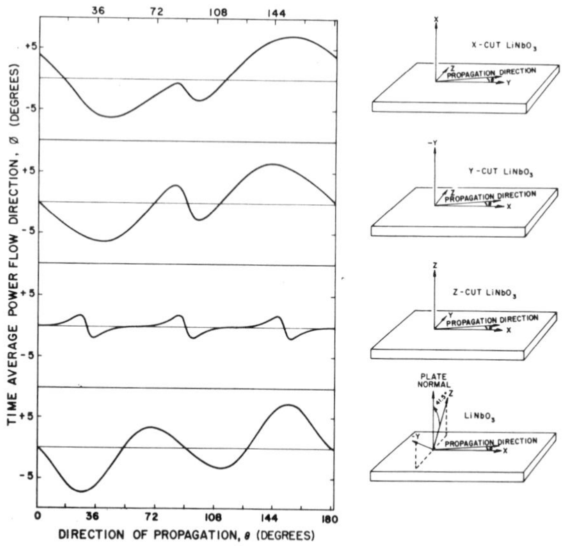

Figure 7 - Angular difference between mechanical time average power flow and phase velocity vector as a function of direction of propagation on four cuts of $LiNbO_3$. The direction of propagation is defined as the direction of the phase velocity.

beam can then miss part or all of the output transducer. Theoretical curves giving the angular separation of the group and phase velocity vectors on four cuts of $LiNbO_3$ are shown in Figure 7. Beam steering was studied by focusing the laser to a spot small with respect to the acoustic beamwidth, and scanning the laser across the acoustic beam. Profiles of surface waves at two frequencies obtained at various distances from the input transducer are presented in Figures 8 and 9. By observing the beam area intercepted by the output transducer one can determine the additional loss caused by beam steering.

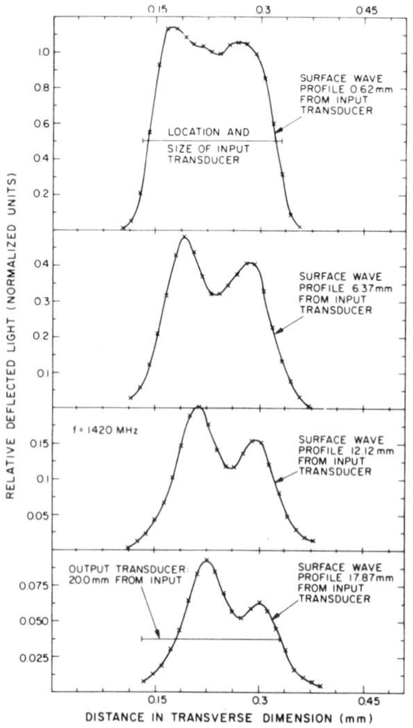

Figure 8 - Laser measurements of acoustic surface wave profiles at various distances from the input transducer. Propagation was along the Z-axis of Y-cut $LiNbO_3$ at a frequency of 1420 MHz. Interdigital transducers were fabricated by electron beam techniques[19] and were operated on their fundamental frequencies.

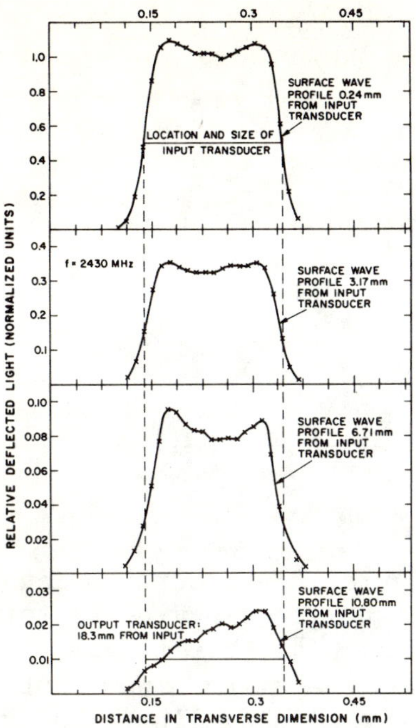

Figure 9 - Laser measurements of acoustic surface wave profiles at various distances from the input transducer. Propagation was along the Z axis of Y-cut $LiNbO_3$ at a frequency of 2430 MHz. Interdigital transducers were fabricated by electron beam techniques[19] and were operated on their fundamental frequencies.

Nonlinear Effects

Let us now return to consider the high power curves shown in Figure 5. In addition to the second-harmonic illustrated there, third-, and fourth-harmonic generation, growth, and decay have also been observed on microwave acoustic $LiNbO_3$ surface wave delay lines as illustrated in Figure 10. These nonlinear effects[10] can be significant in magnitude and can cause measurable changes in the behavior of the fundamental surface wave as shown in Figure 5. For example, an area of negative attenuation or "gain" caused by the rapid return of harmonic power into the fundamental has been observed for the fundamental component of the surface wave. This effect may form the basis of a unique surface-wave amplifier. Growth and decay of the sum and difference surface wave components have also been observed when two input frequencies are used (mixing).

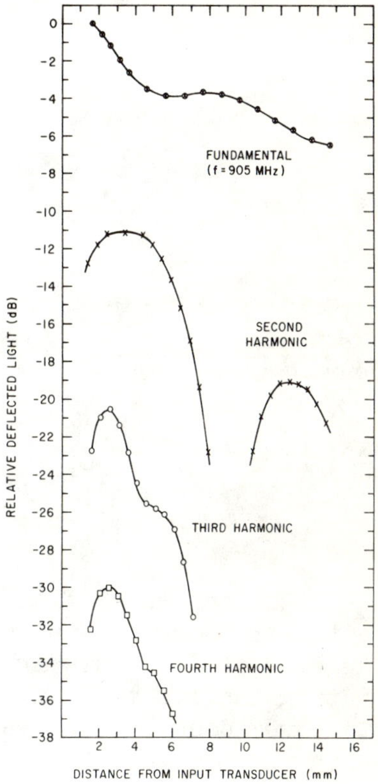

Figure 10 - Deflected light as a function of distance illustrating growth and decay of surface wave harmonics. Fundamental-frequency acoustic power density was 83 mW/mm at the input transducer.

The relative ease with which nonlinear effects can be excited in surface wave delay lines has two important consequences for practical surface wave devices. First, it seems clear that many new devices can be realized, including acoustic harmonic generators, mixers, parametric amplifiers and limiters. The operation of such a limiter is in fact illustrated in Figure 11. On the other hand the observed nonlinearities will definitely limit the power handling capability of other proposed devices, such as delay lines and signal processors. We can show, for example, that at 905 MHz on Y-cut, Z-propagating $LiNbO_3$ nonlinearities become appreciable above 10 mW/mm.

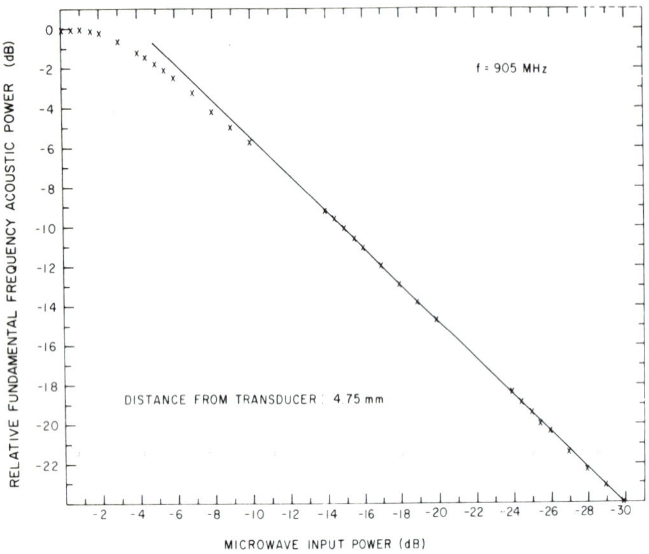

Figure 11 - Light deflected from a 905 MHz fundamental-frequency acoustic surface wave as a function of microwave input power illustrating the limiting effect. The acoustic power does not increase for increases in input power above a certain level. Acoustic power density at the 0 - dB level of input power was 123 mW/mm at the transducer.

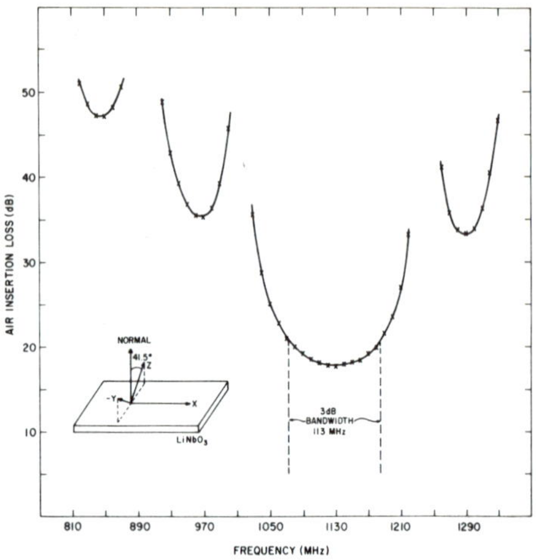

Figure 12 - Insertion loss versus frequency characteristics of a 41.5° rotated-cut, X-propagating $LiNbO_3$ surface wave delay line. No external matching of any kind was used.

A New Cut of $LiNbO_3$ for Surface Wave Applications

The existence of a new high-frequency high-coupling low-beam-steering cut for acoustic-surface-wave propagation on $LiNbO_3$ has been theoretically predicted and experimentally verified.[11] As illustrated in Figure 12, 50 Ω operation with a 10% bandwidth is possible under completely untuned conditions at a frequency of 1130 MHz. This new cut should be particularly useful for relatively inexpensive microwave-frequency delay lines requiring long time delays, since external inductive matching and precision X-ray alignment to minimize beam steering are not as critical as with the YZ orientation.[17, 18]

CONCLUSIONS

By using the laser probe one can account for all sources of insertion loss in surface wave devices. For example, transducer efficiency measurements can easily be made by subtracting propagation and beam steering losses, transducer mismatch losses, and the 6 dB bidirectionality loss from the total electromagnetic-to-electromagnetic insertion loss. This has been done for two YZ $LiNbO_3$ delay lines with the results used to predict the ratio of acoustic power generated on one device to that generated on the other under identical incident power conditions. These results are summarized in Table I. Direct comparison of the light deflected from each of the two surface waves yielded a surface wave power ratio of 1.50 dB. This is in excellent agreement with the value given in Table I.

In conslusion, we see that the laser light deflection technique described in this article is a powerful tool for studying microwave acoustic surface waves. It has been used to obtain design data on nonlinear effects, acoustic propagation losses, beam steering and diffraction losses and insertion loss versus frequency characteristics of interdigital transducers. It is certainly one of the standard methods of investigation in this area.

	Delay Line A (16 mm transducer separation at 960 MHz)	Delay Line B (20 mm transducer separation at 985 MHz)
1. Total insertion loss	23.66 dB	26.45 dB
2. Air propagation loss	3.2 dB/cm = 5.10 dB	3 dB/cm = 6.00 dB
3. Beam-steering loss	4.49 dB	3.85 dB
4. Bidirectionality loss	6.00 dB	6.00 dB
5. Right transducer mismatch loss	1.15 dB	3.39 dB
6. Left transducer mismatch loss	1.61 dB	3.40 dB
7. Computed rf conduction loss per transducer (item 1 minus items 2 through 6)	2.65 dB	1.90 dB
8. Computed ratio of surface-wave acoustic power for equal incident power levels (3.80 dB arbitrary reference level minus items 5 and 7)	0 dB	1.49 dB

Table 1 - Delay Line Characterization by Laser Light Deflection

REFERENCES

1. Special Issue on Microwave Acoustics, IEEE Trans. Microwave Theory and Techniques, MTT-17, (November 1969).

2. Slobodnik, A.J., Jr. (1969) Microwave frequency acoustic surface wave propagation losses in $LiNbO_3$, Appl. Phys.Letts., 14, 94-96.

3. Slobodnik, A.J., Jr. (1970) Microwave acoustic surface wave investigations using laser light deflection, Proc. IEEE 58, 488-490.

4. Richardson, B.A. and Kino, G.S. (1970) Probing of elastic surface waves in piezoelectric media, Appl. Phys.Letts., 16, 82-84.

5. Krokstad, J. (1967) Scattering of light by ultrasonic surface waves in crystals, Norwegian Inst. Tech., Trondheim, Norway, Elab Rpt TE-93 (unpublished).

6. Korpel, A., Lamb, L.J., and Sievering, H.C. (1967) Measurement of acoustic surface wave propagation characteristics by reflected light, Appl.Phys.Letts., 10, 295-297.

7. Auth, D. C. and Mayer, W.G. (1967) Scattering of light reflected from acoustic surface waves in isotropic solids, J.Appl.Phys., 38, 5138-5140.

8. Salzmann, E., Plieninger, T., and Dransfeld, K. (1968) Attenuation of elastic surface waves in quartz at frequencies of 316 MHz and 1047 MHz, Appl.Phys.Letts., 13, 14-15.

9. Slobodnik, A.J., Jr., and Conway, E.D. (1970) The effect of beam steering on the design of microwave acoustic surface wave devices, International Microwave Symposium Digest, 314-318.

10. Slobodnik, A.J., Jr. (1970) Nonlinear effects in microwave acoustic $LiNbO_3$ surface-wave delay lines, JASA, 48, 203-210.

11. Slobodnik, A.J., Jr. and Conway, E.D. (1970) New high-frequency, high-coupling, low-beam-steering cut for acoustic surface waves on $LiNbO_3$, Electronics Letters, 6, 171-173.

12. Bechmann, P. and Spizzichino, A. (1963) The scattering of electromagnetic waves from rough surfaces, New York : MacMillan

13. DeLorenzo, J. D. and Cassedy, E.S. (1966) A study of the mechanism of sea surface scattering, IEEE Trans. Antennas and Propagation, AP-14, 611-620.

14. Shaw, H.J., Reeder, T.M., Gerard, H.M., and Smith, W.R. (1969) Tech.Rpt. RADC-TR-69-79, Stanford University, Stanford, Calif. (unpublished).

15. Slobodnik, A.J., Jr., Carr, P.H., and Budreau, A.J. (1970) Microwave frequency acoustic surface wave loss mechanisms on $LiNbO_3$, Jour.Appl.Phys. 41, 4380-4387.

16. Reeder, R.M., Westbrook, E.M., and Winslow, D.K. (1970) Visualization of surface acoustic waves, Electronics Letts 6 30-32.

17. Campbell, J.J. and Jones, W.R. (1968) A method for estimating optimal crystal cuts and propagation directions for excitation of piezoelectric surface waves, IEEE Trans. Sonics and Ultrasonics, SU-15, 209-217.

18. Collins, J.H., Gerard, H.M., and Shaw, H.J. (1968) High-performance lithium niobate acoustic surface wave transducers and delay lines, Appl.Phys. Letts., 13, 312-313.

19. Broers, A.N., Lean, E.G., and Hatyakis, M. (1969) 1.75-GHz acoustic-surface-wave transducer fabricated by an electron beam, Appl.Phys.Letters, 15 (No. 3) : 98-101.

OPTICAL DEFLECTION AND MODULATION WITH ACOUSTIC SURFACE WAVES IN A NARROW GAP FABRY-PEROT INTERFEROMETER

Ronald L. Lawson
Magnavox Systems, Incorporated
1505 East Main Street
Urbana, Illinois 61801

I. INTRODUCTION

The diffraction of light by bulk acoustic waves in transparent solids and liquids has been studied for over forty years.[1-3] During this period, a number of devices utilizing this effect have been built for the modulation and deflection of light beams. More recently attention has been directed to the diffraction of an incident laser beam by acoustic surface waves.

The scattering of a light beam by acoustic surface waves was first observed in 1966.[4] Since that initial observation, this scattering has been measured both for transmission and reflection of light from a number of materials.[5,6,7]

Following these early studies, this interaction found acceptance as an attractive method of probing surface waves. It allows one to probe various surface points without disturbing the surface wave distribution significantly with commonly used optical intensities. This optical interaction has been used for measurements of surface wave velocity,[8,9] determination of surface wave attenuation,[10,5] measurement of surface wave amplitude,[10] and other studies.

In principle the interaction of light with surface waves can be used to construct devices corresponding to acoustic bulk wave devices for light modulation and deflection. It was this possibility that provided some of the original impetus for research in this area.[10] However, the weakness of this interaction soon made it apparent that surface waves make rather poor deflectors or modulators of light unless some means were found for increasing the effectiveness of the interaction.

In an effort to compensate for the small interaction volume available, the use of a narrow gap optical resonator with one end of the cavity supporting the surface waves has been investigated. This arrangement has been found to greatly increase the amount of light deflected by surface waves. The following sections describe the principles of operation of this system, present experimental results, and summarize some of the advantages and disadvantages of this method for deflection and modulation applications.

II. BACKGROUND

As mentioned in the Introduction, the interaction between acoustic surface waves and an incident light beam is relatively weak. The relative intensity of the nth diffraction order is:[4,7]

$$\frac{I_n}{I_{inc}} = J_n^2\left[\left(\frac{4\pi a}{\lambda}\right)\cos\phi\right]$$

where I_{inc} = incident light intensity

J_n = n^{th} order Bessel function

ϕ = angle of incidence

a = amplitude of the surface wave

and λ = the optical wavelength

For small amplitude surface waves, the expansion of J_1^2 for the first order normalized intensity becomes

$$\frac{I_1}{I_{inc}} \simeq \frac{4\pi a^2}{\lambda^2}$$

for near normal incidence. For a relatively strong surface wave amplitude of 50Å and 6328Å light normally incident, the amount of light diffracted into the first order is about 0.3% of that incident. This amount of light scattered from the incident beam is too small to normally be of interest for practical applications.

If the acoustic surface waves are generated on one end of a narrow gap optical cavity, the scattering efficiency of the acousto-optic interaction can be greatly increased,[11,12] as shown in Figure 1. This figure is a photograph of the Fourier plane of the diffraction pattern with light incident directly on a crystal and the diffraction pattern with the same amplitude acoustic surface waves in a narrow gap resonator. Many more orders are visible with the use of a cavity, and the ratio between adjacent orders is larger.

The basic element in this scheme is a Fabry-Perot interferometer. This device consists of two extremely flat partially transmitting mirrors having a precisely controlled separation. A parallel monochromatic light beam normally incident will be almost completely transmitted through the interferometer after experiencing a large number of multiple reflections when the mirror separation is an integral number of half wavelengths.

The amplitude and phase of the transmitted light is very sensitive to changes in the mirror separation. If one of the interferometer mirrors has small amplitude spatially periodic distortions on its surface, and the mirror separation is set for maximum

A

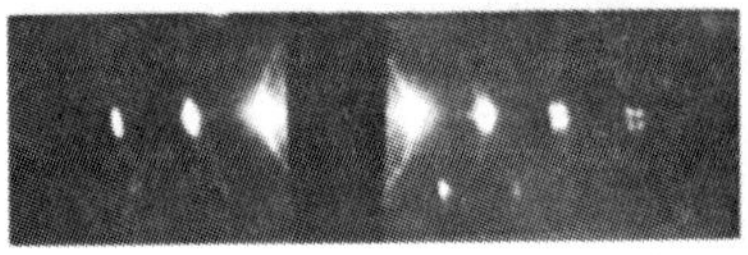

B

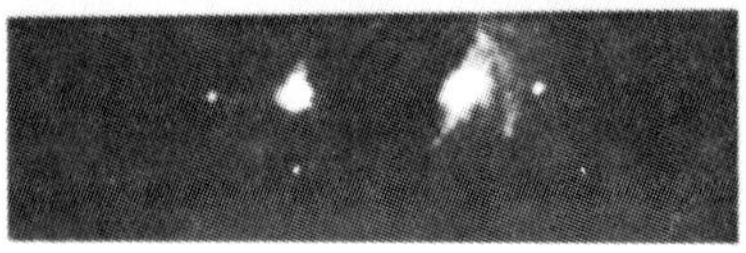

Fig. 1(a) Diffraction pattern produced by transmission through a Fabry-Perot interferometer with surface waves on one mirror (b) Diffraction pattern produced by reflection directly from a mirror with same amplitude waves as (a).

transmission of the incident radiation, the transmitted beam will exhibit periodic amplitude and phase variations corresponding to changes in mirror separation. A complete analysis of the effect of periodic distortion on one mirror of a Fabry-Perot interferometer has been performed by Hunsinger.[13,14]

Basically there are two regions of separation bias at which one may attempt to work. These two regions are shown in Figure 2. One is operation on the step side of the transmission curve and the other is with operation on the peak of the transmission curve.

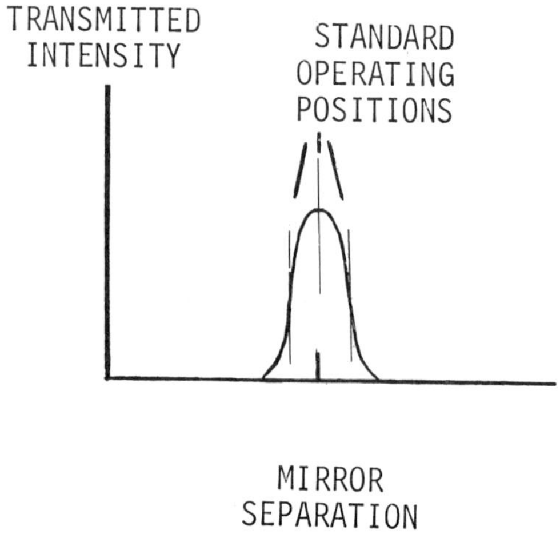

Fig. 2. Standard Mirror Separation

Although operation on the peak does not provide much amplitude modulation with small spatial distortions, it can provide very strong phase modulation. For most modulation and deflection techniques, operation at the peak has been found far more effective.

Surface Wave Generation

Acoustic surface waves provide a means of generating periodic mirror displacements with controllable amplitude and periodicity. These waves have typical amplitudes of a few nanometers with their total energy confined within a few wavelengths of the surface on which they propagate.[15]

Several types of transducers can be used to generate surface waves. One of the most widely used methods involves generation on piezoelectric substrates by piezoelectric coupling with electrode arrays. It is difficult to fabricate piezoelectric mirrors of sufficient flatness and size for use in the cavity.

Therefore, the interferometer mirrors are manufactured of nonpiezoelectric materials, and it is not possible to use electrode arrays to generate surface waves directly using the piezoelectric effect on the mirrors themselves. Figure 3 shows the arrangement for producing surface waves on the mirrors. The waves are generated by the excitation of an interdigital electrode array on the surface of a $LiNbO_3$ crystal and are coupled onto the mirrors through an interface material. The losses involved in the coupling region have been experimentally determined to be less than 4 dB for 20 MHz surface waves and to increase slowly as the frequency rises.

Interdigital arrays can provide a large bandwidth. Broadband operation may be secured by use of variable spacing between adjacent fingers over the length

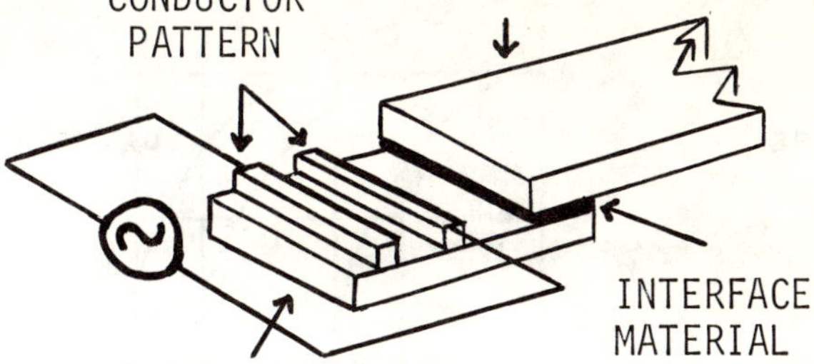

Fig. 3. Coupled Surface Wave Transducer.

of the array. Relatively broadband devices (in absolute terms) may also be obtained by the use of an array consisting of a very few pairs of electrodes centered at a high frequency.

III. LASER BEAM MODULATION

The surface wave grating created in the Fabry-Perot interferometer can be used to modulate the intensity of the transmitted radiation. The surface wave amplitude which is controlled by the information signal strength provided to the transducer determines the amount of light deflected. By absorbing the diffracted light, it is theoretically possible to completely stop the transmitted beam with an electrical signal and achieve 100% modulation depth (controllable intensity range/maximum intensity). Three of the main operating parameters of interest are the modulation power, optical losses, and the maximum data rate.

The modulation power is strongly related to the reflectivity of the mirrors used in the interferometer. The modulation depth is plotted against the surface wave amplitude for various mirror reflectivities below.

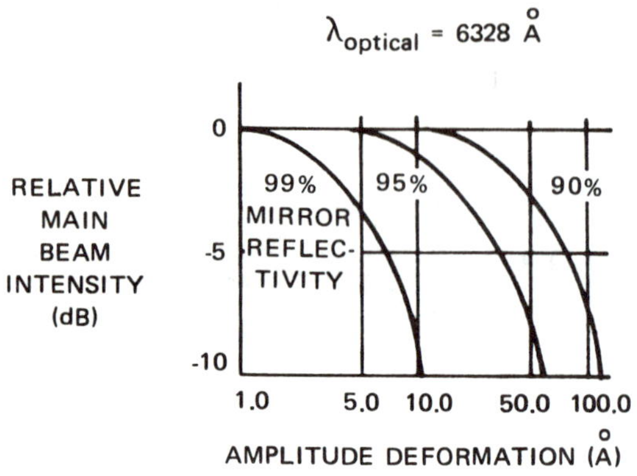

Fig. 4. Effect of Mirror Reflectivity On Modulation Depth As A Function of Surface Wave Amplitude.

The modulation depth increases with mirror reflectivity because of the increased number of multiple reflections in the cavity; consequently, modulators made with higher reflectivity mirrors have smaller modulation power requirements. The upper boundary on mirror reflectivity results primarily from optical losses which also increase as the reflectivity increases. Of the light incident on the mirror, that fraction which is absorbed and scattered is defined as the absorption coefficient. In the visible region, the absorption coefficient can be maintained at less than 0.0005 for reflectivities less than or equal to 99%. Since the choice of mirror reflectivity is a compromise between optical losses and modulation power considerations, nominal values of 99% appear to be most practical.

The modulation power also depends on the mirror material. The acoustic power per unit width of a surface wave beam is related to the peak value of the surface wave amplitude A by:

$$P = K\omega A^2$$

where K is a material constant, and ω is the surface wave radian frequency.

The material constant K defined by this equation is dependent upon the material properties of the mirror. Some typical values of K are:

CdS	Basal Plane	115 wsec/mm^3
Quartz	Y Cut Z Prop	52 wsec/mm^3

Surface wave amplitudes of 3Å are required to produce an 80% modulation depth for 6000Å radiation with 99% reflectivity mirrors.

Modulation power is minimized when the surface wave beam is only as wide as the optical beam d. The effective surface wave power $[P_\varepsilon(\omega)]$ in dBm (that power which actually takes part in the modulation process) as a function of frequency (f) is:

$$P_\varepsilon(\omega) \simeq 10 \log (P_{80}(\omega) \cdot d) + 30)$$

Excessively wide, non-uniform and misaligned surface wave beams result in wasted surface wave power.

The surface wave transducer is a prime consideration in the design of the modulator, because its insertion loss and bandwidth have a large effect on modulation power.

The insertion loss (I_{SW}) of 30% bandwidth surface wave transducers vary from 5 to 50 dB, dependent primarily on the transducer material. The surface wave power increases with frequency; the center frequency must be at least four times the modulation bandwidth (BW_m). The modulation power for a bandwidth BW_M is then:

$$P_m(BW_m) = P_\varepsilon(4 \times BW_M) + I_{SW} + U_{SW}$$

The incident optical beam diameter, d, and the surface wave phase velocity limit the modulation rate, because the surface waves interact with the radiation from the time they enter the beam until they leave it. If the surface wave velocity on the mirror is V_s, the time required for a surface wavefront to cross the beam is d/V_s.

A high frequency wave train of duration d/V_s causes light to be diffracted for a time $2d/V_s$. The maximum bandwidth of a modulator is approximately $V_s/2d$; it occurs when the surface wave train length (d_s) is equal to d.

The size to which the spot can be focused without a beam divergence of more than 0.005 radians and the phase velocity are, thus, primary limitations on the data rates available with such a device.

Experimental Modulation Results

The basic analytic predictions such as modulation depth versus surface wave amplitude and mirror reflectivity have been verified at several mirror reflectivities for 6328Å wavelength light.

One experimental device shown in Figure 5 has achieved a 50% modulation depth of the main beam with an acoustic power density of 33 mw per mm of acoustic wavefront using 95.5% reflectivity mirrors. It has been incorporated into an experimental voice link.

Fig. 5. Photograph of an Experimental Modulator.

In an effort to minimize the acoustic power requirements, recent efforts have involved the incorporation of 99% reflectivity mirrors into the device. Using mirrors of 99% reflectivity, modulation depths in excess of 80% have been achieved using acoustic power densities of less than 20 mw per mm.

The use of 99% reflectivity mirrors demands a very precise control of the spatial separation and mirror parallelism.

Heating effects arising from acoustic losses in the interface medium and the interferometer mirrors cause the interferometer to go out of adjustment relatively quickly when reflectivities of greater than 96% are used. To allow substained operation, an electronic feedback system capable of controlling two tilt axes and the linear separation has been constructed. This system provides continuous readjustment of the cavity. The only effect on modulator operation is to provide a slight <0.5% 60 Hz amplitude modulation on the exiting beam.

No attempts have been made to date to produce modulation at rates in excess of 100,000 Hz, since the present investigations are aimed at a broad study of possible capabilities and not at developing a device used for a specific purpose.

IV. LASER BEAM DEFLECTION

The scattering with this arrangement is of the Raman-Nath type with light diffracting into many side orders rather than into a single order as in Bragg diffraction.

The angular diffraction of the orders depends upon the acoustic wavelength Λ and the optical wavelength λ. For the relatively small angles oridinarily encountered, the angular diffraction of the nth order Θ_n is given by

$$\Theta_n = \frac{n\lambda}{\Lambda}$$

The relative intensity of the various orders is, of course, dependent on the surface wave amplitude and the mirror reflectivities. The figure below shows the variation of zeroth, first, and second orders for a mirror reflectivity of 95.5% as a function of the surface wave amplitude.

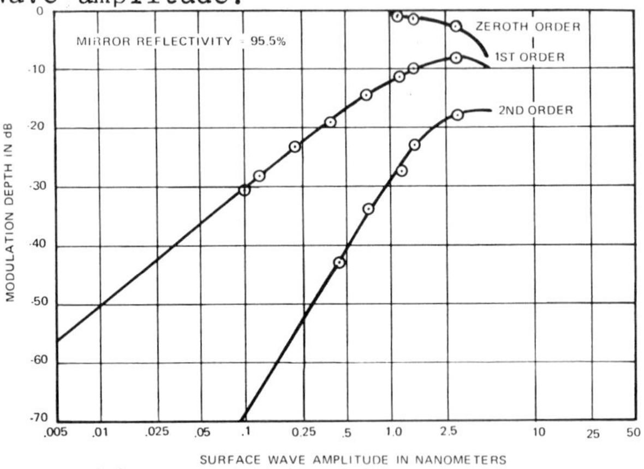

From this graph, it is seen that the intensity of a side order rises as the surface wave amplitude increases until it reaches a maximum value. Further increases in the surface wave amplitude past this point cause a reduction in the intensity of a given order. For scanner applications, maximum intensity in the scanning spot would be obtained by operating at the appropriate amplitude to maximize the side order intensity.

In a scanning system, the important performance parameter is N, the number of resolvable spots to which the scanning beam may be focused.

The effective grating created by the surface waves in the optical cavity causes a portion, I_1, of the incident light beam to be deflected into the first order directions separated from the direction of incidence by the angle $\pm \lambda/\Lambda$ where λ is the optical wavelength and Λ the surface wave wavelength.

If the frequency f_1 for which the acoustic wavelength is Λ_1 is shifted to f_2, the corresponding deflection angles become $\pm \lambda/\Lambda_2$.

If $(f_2 - f_1) = \Delta f$ is the maximum usable bandwidth of the transducer generating the surface waves, then the maximum deflection angle of the system is:

$$\Delta\Theta_{scan} = \lambda/\Lambda_2 = \lambda/\Lambda_1$$

or $$\Delta\Theta_{scan} = \lambda/V\,\Delta f$$

where V is the surface velocity on the interferometer mirrors and $\Delta f = f_2 - f_1$.

The diffracted beam has the same shape as the original beam if the acoustic field fills the aperture. If the incoming beam has a uniform power density over a circular aperture of diameter D and spatial coherence across D, the beam spread will be determined by diffraction. The first nulls of the far-field pattern will be separated from the central maximum by the angle λ/D. Defining this λ/D as the smallest resolvable angle Θ_{min}, the number of resolvable spots N in the first order scan becomes

$$N = \frac{\Delta\Theta_{scan}}{\Theta_{min}} = D/V\Delta f$$

If the desired number of resolvable spots N in the first order scan is 1000, the equation becomes

$$1000 = D/V\Delta f$$

If D is taken as the maximum practical aperture of 2 cm., and V the surface wave velocity is 3000 m/sec., the Δf necessary for 1000 spot resolution is

$$\Delta f = \frac{1000V}{D} = \frac{(1000)(3000\text{m/sec})}{2 \times 10^{-2}\mu}$$

$= 150$ MHz

of the 500 spot resolution using the same parameters $\Delta f = 75$ MHz

If it is desirable to switch the position of the light beam randomly for the purpose of providing rapid access to a specific spot, the maximum switching speed is determined by T_A, the transit time of the acoustic wave across the aperture.

For an aperture of 2 cm. and a velocity of 3000 m/sec

$$T = \frac{D}{V} = \frac{2 \times 10^{-2} \text{ m}}{3 \times 10^3 \text{ m/sec}} = 6.7 \ \mu\text{sec}$$

for an access time of 6.7 microseconds. This corresponds to a frequency of 150 KHz.

It is seen that decreasing the random access time requires a greater acoustic bandwidth to hold the number of resolvable spots constant. This tradeoff is unavoidable and involves compromises depending on the end use of the deflector.

With the use of Raman-Nath scattering for wideband deflection systems, care must be taken to prevent overlap of the range of adjacent side orders unless the intensity dropoff to the next higher order is sufficient that overlap of orders may be neglected. For 95.5% reflectivity mirrors, the maximum intensity in the first order occurs when the second order intensity is 10 dB below first order intensity. This is not a sufficient dropoff for neglecting the second order in most applications. To overcome the overlap problem, the amplitude of the surface waves may be reduced until the second order is sufficiently far below the first order. This practice is undesirable because it also greatly reduces the amount of light deflected in the first order.

The second method of avoiding overlap involves choosing the center frequency f_0 such that for a desired half bandwidth $\Delta f/2$

$$f_0 + \frac{\Delta f}{2} < 2(f_0 - \frac{\Delta f}{\sigma})$$

or

$$f_0 > \frac{3\Delta f}{2} .$$

Power Requirements

Maximum diffraction of light into the first orders occurs at an amplitude of approximately 0.5 nanometers for 99% reflectivity mirrors.

The acoustic power P_A per millimeter of beamwidth of an acoustic surface wave disturbance can be written in terms of a material constant (K) and the wave amplitude (A_s) and the radian frequency (ω) as follows:

$$P_A = K \ A_s^2 \ \omega$$

The value of K varies from material to material. Some typical values are

The value of K for the interferometer mirrors presently being used is somewhat less than 100 watt-sec/mm^3. Assuming K = 100, the acoustic power requirements for an amplitude of 0.5 nanometers at 100 MHz is:

$$P_A = 15.7 \times 10^{-3} \ \frac{\text{watts}}{\text{mm}}$$

of beamwidth.

The true drive power required to produce this acoustic power density of the interferometer loss and the additional losses at the coupling interface. This last loss factor could be eliminated if one of the interferometer mirrors were to be constructed from a strongly piezoelectric material. For a given bandwidth, Δf, the minimum center frequency necessary to satisfy the condition of no order overlap is $3\Delta f/2$. The total power requirements per millimeter of beamwidth at the center frequency becomes:

$P = (25 \times 10^{-12})(3\Delta f)$ watt/mm $+ I_{SW}$

$+ C_{SW}$

or $P = (23.5 \times 10^{-11})(\Delta f)$ watts/mm

where I_{SW} = transducer insertion loss

and C_{SW} = acoustic loss in coupling.

Experimental Deflection Results

Figure 6 is an oscillogram of the intensity variation of the first order as the driving frequency is swept. This particular photo was taken using a $LiNbO_3$ crystal transducer with four pairs of interdigital grids and an 80 MHz center frequency. In this case, no attempt was made at matching the transducer to its driver.

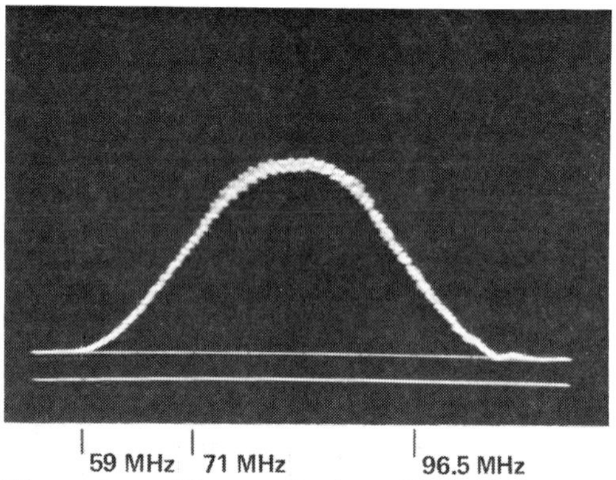

Fig. 6. First Order Intensity vs. Frequency

The diffracted light intensity is seen to vary greatly as a function of frequency.

Further experiments not yet complete indicate that a 30 MHz flat sweep centered at 80 MHz should be feasible. Because this is a Raman-Nath type process with this arrangement, it is not possible to scatter with high efficiency into a single order, and this is a definite disadvantage for applications.

V. CONCLUSION

The use of a Fabry-Perot interferometer has been investigated as a means of increasing the scattering efficiency of acoustic surface waves. The scattering efficiency of given amplitude surface waves can be increased by a factor of several hundred if high reflectivity mirrors are used for the Fabry-Perot cavity.

This method of diffraction can be used for constructing light deflection and modulation systems and combination deflection modulation devices for laser displays. However, this technique has several disadvantages for practical usage.

The Fabry-Perot interferometer with high reflectivity mirrors is a very sensitive device. As such, its transmission efficiency is very sensitive to vibration, noise, and thermal fluctuations. Isolation from these factors is difficult to obtain in other than a laboratory situation. Failure to isolate the device from these factors may result in periods of signal blackout and wide amplitude variation of the output.

The relatively slow phase velocity of the acoustic waves, while an advantage in many applications, makes it difficult to modulate at rates in excess of several hundred megabits per second due to the finite transit time across the incoming optical beam.

Additionally this is a multiple order diffraction process and the amount of light available in a single order is limited. For deflection application and display systems, this is undesirable.

With the rapid strides being made with electro-optic devices and the present state of development of bulk acousto-optic devices, it does not appear that this technique for enhancing surface wave diffraction is likely to be of practical significance in laser beam deflection and modulation.

This enhanced diffraction technique may be applicable to certain types of signal processing, especially two dimensional processing applications and have useful consequences for certain holographic applications.

Investigation of these applications has not advanced far enough to determine their feasibility and practical value.

REFERENCES

1. L. Brillouin, "Diffusion de la lumiere et des rayons X par un corps transparent homogene," Ann. Phys. (France), 9th series., Vol. 17, p. 88, 1922.

2. P. Debye and F. W. Sears, "On the scattering of light by supersonic waves," Proc. Nat. Acad. Sci. (USA), Vol. 18, p. 409, 1932.

3. M. Born and E. Wolf, Principles of Optics, New York: Pergamon, 1959.

4. E. P. Ippen, "Diffraction of light by surface acoustic waves," Proc. IEEE, Vol. 55, No. 2, pp. 248-249, February, 1967.

5. A. Korpel, L. J. Laub, H. C. Sievering "Measurement of acoustic surface wave propagation characteristics by reflected light," Appl. Phys. Letters, Vol. 10, No. 10, pp. 295-297, May 15, 1967.

6. J. Krokstad and L. O. Svaasand, "Scattering of light by ultrasonic surface waves in quartz," Appl. Phys. Letters, Vol. 11, No. 5, pp. 155-157, September 1, 1967.

7. D. C. Auth and W. G. Mayer, "Scattering of light reflected from acoustic surface waves in isotropic solids," Jour. Appl. Phys., Vol. 38, No. 13, pp. 5138-5140, December, 1967.

8. F. W. Voltmer, E. P. Ippen, R. M. White, T. C. Lim, and G. W. Farnell, "Measured and calculated surface-wave velocities," Proc. IEEE, Vol. 56, pp. 1634-1635, September, 1968.

9. B. H. Soffer, D. H. Close, and M. E. Pedinoff, "An optical imaging method for direct observation and study of acoustic surface waves," Appl. Phys. Letters, Vol. 15, pp. 339-342, November 15, 1969.

10. E. P. Ippen, "Diffraction of light by acoustic surface waves," PhD dissertation, University of California, Berkeley, 1968.

11. B. J. Hunsinger and D. Holshouser, "Spation modulation of light using surface waves in an interferometer," Appl. Phys. Letters, Vol. 16, No. 7, April 1, 1970.

12. R. L. Lawson, "A new electronically controlled acousto-optic laser beam modulator," Presented at the 1970 Government Microcircuit Applications Conference, October 6-8, 1970, Ft. Monmouth, New Jersey.

13. B. J. Hunsinger, "Analysis of an interferometer with acoustic surface waves on one mirror," Appl. Optics, Vol. 10, No. 2, February, 1971.

14. B. J. Hunsinger, "Spatial modulation of light using surface waves in a Fabry-Perot interferometer," PhD dissertation, University of Illinois, Urbana, 1970.

15. R. M. White, "Surface elastic waves," Proc. IEEE, Vol. 58, No. 8, August, 1970.

A NOTE ON THE PERTURBATION THEORY FOR PIEZOELECTRIC COUPLING TO SURFACE WAVES

H. Hanebrekke, K.A. Ingebrigtsen and L. Solie
Division of Physical Electronics
The Norwegian Institute of Technology
7034 TRONDHEIM, Norway

Recently, Skeie [1] and Lakin [2] have developed perturbation theories which describe the electromagnetic coupling to surface acoustic waves. Skeie's approach, which also accounts for a mechanical loading of the surface, is based on an acoustic and electric impedance wall concept developed by Ingebrigtsen and Tonning [3,4], whereas Lakin's theory is closely related to the normal mode theory developed by Auld [5]. Both approaches lead to equivalent results in the first order approximation.

The theories describe the perturbation in wave number (or phase velocity) that results from a change in the electromagnetic properties of the adjoining medium. A basic assumption in the first order approximation of both approaches is that the mechanical stress and the total surface wave power remain constant under the perturbation. Comparison with exact calculations [6] shows very good agreement for several important cuts and orientations of $LiNbO_3$ and $Bi_{12}GeO_{20}$, whereas in other cases, for example the Y-cut X-oriented $LiNbO_3$, the agreement is rather poor. Lakin also points out that his approach cannot be applied to the piezoelectric SH-surface wave described by Bleustein [7].

It is the purpose of this note to show that Skeie's or Lakin's approach with a small modification may be extended to cover the case of (YX)-$LiNbO_3$, as well as the case of (XY)-$LiNbO_3$ which shows a similar disagreement in first order perturbation. The particular case of the piezoelectric SH-wave will not be considered since this case may be given an exact analytical treatment [8].

In the following we shall use Lakin's approach since it starts with an exact expression for the wave number perturbation as a basis for the expansion.

Assuming wave propagation in the x_3-direction with the x_2-direction normal to the surface, the general expression for the perturbation in wave number k may be written [2].

$$(k-k') \int_{-\infty}^{0} [v_j{}^*T'_{3j} + v_j{}'T_{3j}{}^* + i\omega(\Phi^*D_3{}' - \Phi'D_3{}^*)]dx_2$$

$$= -\omega \int_{-\infty}^{0} \frac{\partial}{\partial x_2} (\Phi^*D_2{}' - \Phi'D_2{}^*)dx_2 \cdot \qquad (1)$$

T_{3j} , v_j , Φ , D_3 , and D_2 are respectively the traction in a plane normal to the x_3-direction, the particle velocity, the electric potential, and the x_3 and x_2 components of the electric displacement. In deriving this expression it is assumed that the surface $x_2=0$ is a traction free surface.

Equation (1) is exactly valid for any pair of solutions denoted by primed and unprimed quantities, respectively. It is worthwhile to notice that when primed and unprimed quantities refer to the same solution, the integral on the left hand side of the equation gives 4 times the negative of the total power transported by the surface wave inside the piezoelectric medium.

Consider two particular cases. Assume first that the primed solution corresponds to a metallized surface. Then

$$\Phi' = 0 \qquad x_2 \leq 0$$

and the right hand side of Eq. (1) may be integrated. The result is

$$(k-k_a) \int_{-\infty}^{0} [v_j{}^*T^a{}_{3j} + v^a{}_jT_{3j}{}^* + i\omega(\Phi^*D^a{}_3 - \Phi^aD_3{}^*)]dx_2$$

$$= -\omega\Phi^*D^a{}_2\Big|_{x_2=0} \qquad (2)$$

where the index "a" refers to the case of a metallized surface.

As the opposite extreme case we may assume

$$D_2 = 0 \qquad x_2 \leq 0$$

which is a hypothetical boundary condition corresponding to zero dielectric constant of the adjacent medium. Denoting this solution with an index "b" we obtain

$$(k-k_b) \int_{-\infty}^{0} [v_j{}^*T^b{}_{3j} + v^b{}_j T_{3j}{}^* + i\omega(\Phi^* D^b{}_3 - \Phi^b D_3{}^*)]dx_2$$

$$= \omega\Phi^b D_2{}^* \Big|_{x_2=0} \qquad (3)$$

The normal component of D may be found from the constitutive relation using the quasi static approximation for the electric field.

$$\varepsilon^T{}_{ij} \frac{d\Phi}{dx_j} = - D_i + \partial_{iI}T_I \qquad (4)$$

where $\varepsilon^T{}_{ij}$ is the permittivity tensor at constant stress and the d_{iI} the piezoelectric strain tensor. Upon taking the divergence of both sides of the equation, the term containing D_i vanishes. In the limit of weak piezoelectric coupling this equation may be solved for the potential by expressing it as the sum of two terms. One of these is the inhomogeneous solution of Poisson's equation with the polarization term $d_{iI}T_I$ acting as an independent forcing term while the other is the homogeneous solution found by setting $d_{iI}T_I$ equal to zero.

As a first order approximation Skeie [1] and Lakin [2] then assume that the mechanical stress and power flow remain constant under a perturbation of the electrical boundary conditions. Further, the change in Φ is assumed to consist only of a change in the amplitude of the homogeneous solution. Since the electric displacement corresponding to this solution is related to the potential by the constant stress dielectric tensor, it is concluded [1,2] that the "effective dielectric constant" should be derived from the constant stress dielectric tensor. This leads to the following expression for $D_2{}^*$ at $x_2=0$.

$$D_2{}^* \Big|_{x_2=0} = - (\varepsilon^T{}_{22}\, \varepsilon^T{}_{33} - \varepsilon^T{}_{23})^{\frac{1}{2}}\, k_b (\Phi^* - \Phi^{*b}) \Big|_{x_2=0} \qquad (5)$$

The perturbation in wavenumber is finally determined by solving Laplace's equation for the adjacent medium and requiring continuity in D_2 and Φ at the interface.

Equation (5) is a valid expression for the normal D field for weakly piezoelectric materials as is confirmed by the success of perturbation theory for these cases, so any alternative expression for D_2 must be consistent with Eq. (5) in this limit. The alternative expression proposed in this paper involves a generalization of the "effective dielectric constant"

$$(\varepsilon^T_{22}\,\varepsilon^T_{33} - \varepsilon^{T2}_{23})^{\frac{1}{2}}$$

to an expression that will be valid not only in the weakly piezoelectric limit.

The point of departure with Eq. (5) comes in the assumption that Eq. (4) can be solved for Φ by assuming the $\bar{d}_{iI}\bar{T}_I$ is a constant forcing term independent of Φ. This assumption leads to the conclusion that the decay constant of Φ in the x_2 direction should be equal to k (when $\varepsilon_{22} = \varepsilon_{33}$) and that the contribution to Φ from the quasiacoustic partial waves can be neglected or at least assumed to be unchanged under a perturbation of the electrical boundary conditions. Exact numerical calculations show that for weakly piezoelectric materials the decay constant of the quasi electromagnetic partial waves is equal to the wavenumber to within a few percent or less (the relative difference in these quantities for GaAs, InAs, InSb are less than 1.1%, 0.07% and 0.3% respectively). However, for

$$\text{YX } LiNbO_3 \; (\varepsilon^T_{xx} = \varepsilon^T_{yy} = 84\,\varepsilon_o)$$

the decay constant of the quasi electromagnetic partial wave differs from the wavenumber by over 66% and the contribution to Φ from the quasi electromagnetic partial wave is only slightly larger than from each of the quasiacoustic partial waves. Thus, it is impossible to distinguish the partial waves making up a surface wave as quasi electromagnetic or quasi acoustic for strongly piezoelectric cases, and to neglect the contribution to Φ from the quasi acoustic partial waves would be invalid.

Unfortunately, it appears that for the strong piezoelectric materials Φ and D can only be found satisfactorily by an exact numerical calculation. This, however, is not a serious drawback since the same field quantities which must be used for the perturbation analysis of the acoustic amplifier are used to determine the "effective dielectric constant".

Equation (2) may be used as the basis for a Taylor expansion of the potential Φ^* in terms of $(k-k_a)$. The first term in the expansion will be simply

$$-\omega\Phi^* D^a_2 \Big|_{x_2=0} \simeq -4(k-k_a)P_a \qquad (6)$$

where P_a is the power transported by the surface wave with wavenumber k_a.

A similar expansion of Eq. (3) gives

$$\omega\Phi^b D_2^* \Big|_{x_2=0} \simeq -4(k-k_b)P_b \qquad (7)$$

where P_b is the power for this case.

When both power integrals are equal, Eqs. (5) and (7) may be combined to eliminate k and express D_2 in terms of D^a_2 and Φ^b. The result is

$$D_2^* \Big|_{x_2=0} = \frac{D^a_2}{\Phi^b} (\Phi^* - \Phi^{b*}) \Big|_{x_2=0} \qquad (8)$$

When we compare this expression with Eq. (5) we find that we may define an "effective dielectric constant" which is

$$\bar{\varepsilon}_{eff} = -\frac{D^a_2}{k_b\Phi^b} \Big|_{x_2=0} \qquad (9)$$

Thus the effective dielectric constant may be expressed as the ratio of the normal component of the electric displacement with a metallized surface and the potential fo an "ideal open" surface when the amplitudes are normalized to give the same flow of power for the two cases. It is worthwhile to note that by dividing Eq. (6) by Eq. (7) we find

$$\frac{\Phi}{D_2} = \frac{\Phi^b}{D_2^a} \times \frac{k-k_o}{k_b-k} \qquad (10)$$

which is equivalent to a relation derived elsewhere from a different approach. (4).

By comparing with the Eqs. (5) - (7) we see that Eq. (8) automatically ensures that the extreme cases for which $k = k_a$ and $k = k_b$ will be exactly correct in the first order perturbation.

In Table I we have given the numerically determined values of $(\varepsilon_{ff}/\varepsilon_o)$ as well as the zero stress permittivities for important cuts and orientations of $LiNbO_3$. It is observed that there is a large difference between the two alternative

Cut/orientation	XY	XZ	YX	YZ	ZX	ZY
$\frac{\varepsilon_{eff}}{\varepsilon_o}$	145.5	54.2	174	52.5	57.9	49.2
$\frac{\varepsilon^T}{\varepsilon_o}$	84	50.2	84	50.2	50.2	50.2

Table I. Numerically determined values of the effective permittivity for various cuts and orientations of $LiNbO_3$. The zero stress values are given for comparison.

expressions for Y-cut X-propagation and for X-cut Y-propagation.

In order to test the expressions we shall use the perturbation theory to determine the phase velocity with an adjacent metallized surface separated by an air gap h from the piezoelectric. From the perturbation theory the relation is given by [2,4]

$$\frac{v - v_b}{v_a - v_b} = \left(1 + \frac{\varepsilon_{eff}}{\varepsilon_o} \tanh kh\right)^{-1} \tag{11}$$

This relation is plotted in Fig. 1 for YX and XY $LiNbO_3$ together with exact calculations. Within the accuracy of the drawing it is not possible to separate the exact calculations from the approximate expression when ε_{ff} from Table I is used whereas the use of ε^T leads to considerable derivation from the exact calculations for these cases. For the other cases given in Table I, the two alternative definitions of the dielectric constant agree fairly well. The agreement with exact calculations for these cases has been demonstrated by Lakin [2].

In conclusion we may say that the use of an effective dielectric constant as defined by Eq. (9) rather than the usual assumption of the zero stress dielectric constant increases the range of validity of the first order perturbation theory. In order to apply the perturbation theory three quantities must be determined numerically, the phase velocity for a shorted surface v_a, the phase velocity for an ideal open surface v_b, and the effective dielectric constant as defined by Eq. (9).

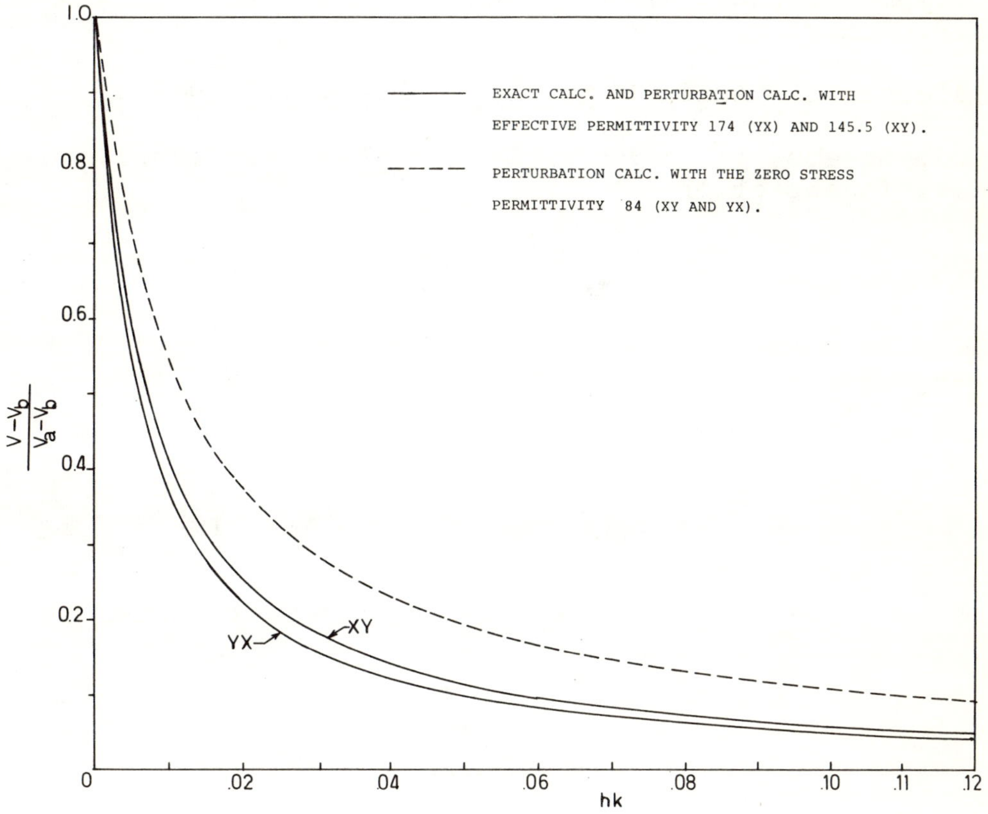

Fig. 1. Variation of phase velocity with distance h to shorting plane (in units of kh) for YX- and XY-$LiNbO_3$. Exact calculations agree within accuracy of the drawing with the perturbation expression using the values of $\varepsilon_{eff}/\varepsilon_o$ from Table I.

LIST OF REFERENCES:

1. H. Skeie: J. Acoust. Soc. Am. 48, (Part 2), 1098 - 1109, (1970).

2. K.M. Lakin: J. Appl. Phys. 42, 899 - 906, (1971).

3. K.A. Ingebrigtsen, and A. Tonning: Phys. Rev. 184, 942 - 951, (1969).

4. K.A. Ingebrigtsen: J. Appl. Phys. 40, 2681 - 2686, (1969).

5. B.A. Auld: IEEE Trans. Microwave Theory Tech. MTT 17, 800 - 811, (1969).

6. J.J. Campbell, and W.R. Jones: IEEE Trans. Sonics Ultrasonics 15, 209 - 217, (1968).

7. J.L. Bleustein: Appl. Phys. Lett. 13, 412, (1969).

8. C.A.A.J. Greebe, and K.A. Ingebrigtsen: Phys. Lett. 30A, 364, (1969).

ACOUSTO-OPTICS

THE STATE OF THE ART IN BEAM STEERING AND CONTROL USING ACOUSTO-OPTICS

H. A. Heynau
Control Laser-Orlando, Inc.
11222 Astronaut Boulevard
Orlando, Florida 32809

INTRODUCTION

Although acousto-optic beam steering and control has been in existence since the early 1930's, it has been the advent of the laser which has rekindled interest in this class of devices. Also, the newer and more sophisticated demands made by state of the art electro-optic systems require fast, electronically controllable means for steering coherent optical beams. Acousto-optic techniques have outpaced electro-optic ones due to the relatively large angles and number of resolvable spots available with present day materials and fabrication techniques. Many papers have been written on various materials and/or specific devices. It is the purpose of this article to bring together the background material, a descriptive materials summary, and a discussion of the systems considerations which bear upon and characterize present day acousto-optic beam deflectors. To provide a background for comparing materials and discussing systems considerations, a brief discussion of acousto-optic background material is in order.

BACKGROUND

The acousto-optic diffraction process is diagrammed in Fig. 1. The ultrasonic energy propagating through the medium forms a three dimensional grating by virtue of the index of refraction variations associated with the traveling acoustic wave. The grating periodicity is equal to the ultrasonic wavelength Λ:

$$\Lambda = \frac{V_s}{f_o}$$

where V_s is the velocity of sound in the medium and f_0 is the frequency being propagated. If the three dimensional grating is thin in the direction of light propagation $\ell < (\Lambda^2/\lambda)$, the incident optical beam will be coupled into several symmetrical diffraction orders--just as in a two-dimensional diffraction grating. Similarly, the orders are separated in angle by approximately λ/Λ where λ is the wavelength of the light beam in the diffracting medium.

As the thickness of the medium, ℓ, is increased in the

direction of light propagation, the amount of light coupled into one of the diffracted orders will increase at the expense of the other diffracted orders. This enhancement is similar to the reflection of x-rays from crystal lattice planes and, for this reason, has taken the name Bragg reflection with the diffraction angle θ_B being given by

$$\sin \theta_B = \tfrac{1}{2}\lambda/\Lambda,$$

called the Bragg angle. This Bragg mode of operation increases the efficiency of coupling power into a diffracted beam and narrows the angular range over which the cell may be used.

As indicated in Fig. 1, the angle between the undeflected and the diffracted beam is $2\theta_B$ which, for the small angles normally involved, is given by

$$2\theta_B = \lambda f/V_s$$

It is this linear dependence of angle on frequency that is one of the acousto-optic deflector's attractive features. By using a voltage controlled oscillator to drive the deflector, an analog, or digital, applied voltage will control the position of the deflected spot. Thus for large screen TV displays, holographic computer memory access, training displays, the VCO is programmed directly with the positional information.

In specifying or designing an acousto-optic deflector there are three main parameters: (1) number of resolvable spots, N, (2) random access time, τ, and (3) deflection efficiency, η. The number of resolvable spots, N, is generally specified in terms of the Rayleigh criterion for the resolution of two diffraction patterns. The problem originally arose in connection with the angular resolution of a telescope or microscope. In the deflector case, the concern is with the ratio of the angular scan capability of the deflector to the minimum resolvable angle--which is limited by the deflector's finite aperture. For reference, Fig. 2 displays various diffraction patterns, with the variable being the angular separation of two sources relative to the Rayleigh criterion for the aperture through which their light has passed. These patterns can be used by system designers as a guide in the selection of the number of Rayleigh criterion resolvable spots required in order to yield some other number of separate spots based on esthetic, system logic level, or cross talk considerations. To find the angular ratio, the total angular scan capability of a deflector is found to be

$$\Delta(2\theta_B) = \Delta\left(\frac{\lambda f}{V_s}\right) = \frac{\lambda}{V_s}\Delta f$$

where Δf is the optical bandwidth of the deflector--which takes into account electronic, transducer and Bragg angle bandwidth considerations. For a light beam of uniform intensity and coherence across an aperture, d, the smallest resolvable angle,

θ_{min}, by the Rayleigh criterion will be

$$\theta_{min} = \lambda/d$$

The number of resolvable spots will therefore be

$$N = \frac{\Delta(2\theta_B)}{\theta_{min}} = \frac{d}{V_s}\,\Delta f = \tau\Delta f$$

where τ is the random access time of the device. Thus the time-bandwidth product, $\tau\Delta f$, is the yardstick by which deflectors and acousto-optic signal processors are compared.

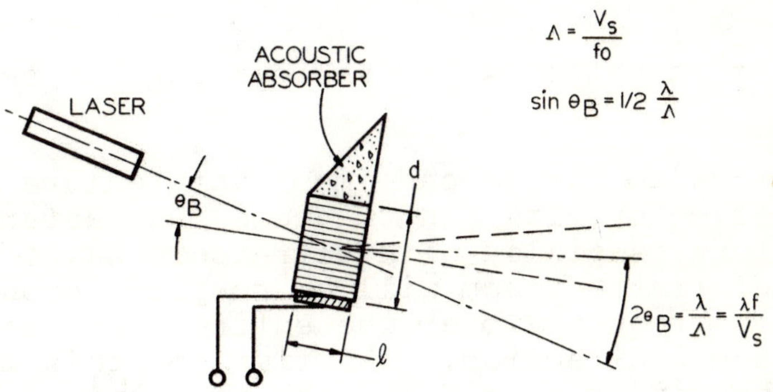

Fig. 1. Acousto-optic diffraction process.

The bandwidth, in this relation, is a result of both the electrical Q of the transducer capacitance-radiation resistance combination and the mechanical Q, which is a result of bonding transducers of one acoustic impedance onto a medium of another acoustic impedance. The fractional bandwidth that can be achieved about a given center frequency is somewhat over 50%.

Another factor that may enter into the bandwidth of a deflector is the bandwidth associated with the Bragg angle. This bandwidth may be estimated to be $\pm\frac{1}{2}\Lambda/l$.[1] This limits the length, l, of the transducer for a given center frequency. This limitation can be expanded somewhat by the use of acoustic beam steering[1] --much as beam steering is used in phased array radars.

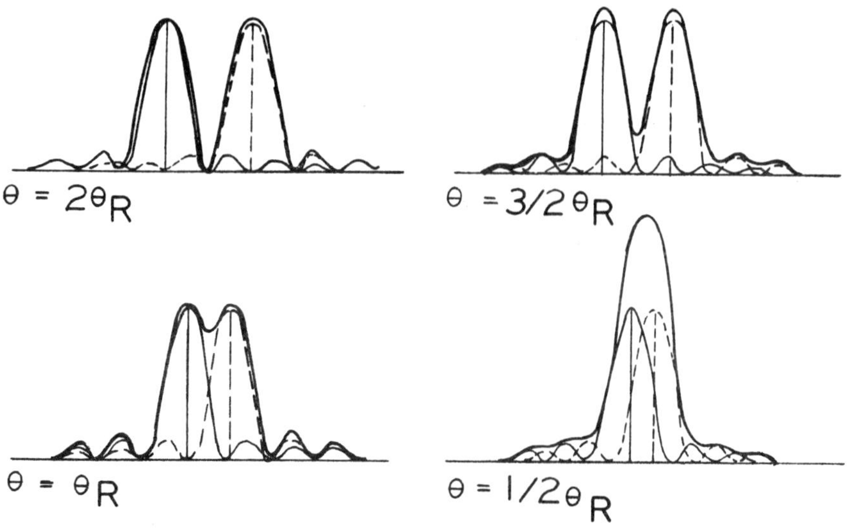

Fig. 2. Diffraction patterns for various angular separations.

This brings us to the second important feature of the acousto-optic deflector--its random access characteristic. When the entire aperture is filled with ultrasonic waves of a single frequency, the diffracted beam will be coupled in one direction. A change to a new frequency over the entire aperture will result in coupling to a new direction. The time for this change to occur is just the time required for the new sonic frequency information to propagate across the cell aperture. This access time, τ, is equal to the dimension of the laser beam in the direction of sound propagation, d, divided by the velocity of sound in the medium, V_s. Thus

$$\tau = \frac{d}{V_s}$$

is the access time and is independent of whether the diffracted beam angle is changed by a small or large amount.

The third main parameter for acousto-optic deflectors, that of deflection efficiency, returns to the diffraction process and the efficiency of coupling optical power from the input beam into one diffracted order. This efficiency is usually defined as the fraction of the incident laser beam that is deflected for one watt of electrical drive power input. Factors affecting this measure of performance are the piezoelectric coupling efficiency of the transducer, the insertion loss of any acoustic matching or bandwidth broadening film layers, and the efficiency of the acousto-optic scattering in the medium selected. To compute the latter quantity, Gordon[2] has shown that the fraction of light power scattered is given by

$$I/I_0 = \sin^2 (\eta)^{\frac{1}{2}},$$

where the parameter η is defined by

$$\eta = \tfrac{1}{2}\pi^2 \left(\frac{n^6 p^2}{\rho v_s^3}\right) (\lambda_o^2 \; H \cos^2\theta_B)^{-1} W_o P_a$$

The variables in this equation may be identified as follows: n = index of refraction of medium, p = photoelastic coupling constant, ρ = density, λ_o = wavelength of light used, H = acoustic beam height, θ_B = Bragg angle, W_o = acoustic beam width, and P_a = total acoustic power.

For an acousto-optic coupling efficiency of 50% at a wavelength of 5145 Å, a simplified expression is

$$P_{50} = 21.8 \, \frac{h}{\ell} \, (M_{2R})^{-1},$$

where P_{50} is the acoustic power required for coupling 50% of the optical beam, h is the height of the sound column in the direction normal to the optical beam propagation direction, ℓ is the length of the acoustic column parallel to the optical propagation direction, and M_{2R} is a normalized figure of merit for the medium to be used.

In using this equation, the figure of merit (M_{2R}) is the conventional M_2 value normalized to the maximum value for fused quartz. Values for M_{2R} for various materials of interest for acousto-optic deflectors as well as other pertinent characteristics of these materials will be discussed in the next section.

Both the insertion loss of changing rf to acoustic energy and the bandwidth of the deflector are governed by the choice of transducer material and bonding technique. Typical low frequency deflectors use ceramic transducers fabricated from one of the Titanate materials and are bonded using epoxy techniques. High frequency devices, up to several hundred MHz, typically use $LiNbO_3$[3] which is bonded means of an indium, cold weld, metallic bond.[4] Indium bonding is used because of its low loss properties and the ability to form thinner bonds than with epoxy techniques. The latter is important in the fabrication of high frequency transducers which large bandwidths make necessary.

MATERIALS

The state-of-the-art in acousto-optic deflection is largely one of materials. That is, a state-of-the-art deflector is one for which the optimum material is selected for the particular application and fabrication and assembly techniques are employed which best achieve the full capabilities of that material. Since material selection involves many tradeoffs which are not always clear in the end user's mind, this section is an attempt to discuss several of the prominent acousto-optic materials and indicate their advantages, disadvantages, and devices in which they might be used.

A. Fused Silica, $n = 1.45$, $V_S = 5.95 \times 10^5$ cm/sec., $M_{2R} = 1.0$

Fused silica is the "bench-mark" material by which other materials are judged. In the past, H_2O was used for this purpose, but since fused silica has a relatively low figure of merit (M_2), it is convenient to take the factors by which other materials exceed that of fused silica rather than the ratio that a material is below that of H_2O, or, db compared to H_2O. Since fused silica has the lowest figure of merit of the commonly used materials, it may seem strange to consider it when discussing state-of-the-art devices. It appears here not only because of its reference status, but because it is available in large pieces of high optical quality. Thus when very large numbers of resolvable spots (N) are required, in the order of 1,000, fused silica becomes one of the prime candidates in spite of its relatively poor deflection efficiency of 3-10%. This material can be used with carrier frequencies up to 350 MHz before acoustic attenuation becomes a factor in most devices.

B. H_2O, $n = 1.33$, $V_s = 1.5 \times 10^5$ cm/sec., $M_{2R} = 106$

As mentioned above, this material has been used as a reference material for judging relative efficiency of acousto-optic coupling of other materials. Relative to fused silica, H_2O has a figure of merit of 106. Even with this high figure of merit, however, this material finds only limited usage due to its relatively high acoustic attenuation. This limitation precludes its use above approximately 40 MHz. To achieve large bandwidths at this low frequency is fairly difficult, and then usually only with the assistance of acoustic beam steering to overcome the Bragg angle "Q" of these devices. See reference 1 for an example of a cell of this type. This medium or the one following are generally used in circumstances where low cost in high volume is the primary consideration.

C. Glasses, $n = 1.7\text{-}1.9$, $V_s = 3\text{-}4 \times 10^5$ cm/sec., $M_{2R} = 5\text{-}15$

Although not as limited by acoustic attenuation as H_2O, most glasses are not suitable for large bandwidth, short access time deflectors and are primarily used in low cost, mass produced devices where large aperture sizes are tolerable and the availability of large pieces is an asset.

D. Alpha-Iodic Acid, $n = 1.93$, $V_s = 2.44 \times 10^5$ cm/sec., $M_{2R} = 55$

Alpha-iodic acid[5] (α-HIO_3) has one of the highest figures of merit of materials suitable for visible wavelength operation. Its acoustic attenuation is low up to approximately 500 MHz (typical loss = 2.5 db/cm @ 500 MHz) which also makes it desirable as a deflector cell material. It does, however, have several practical disadvantages. Alpha-iodic acid is a solution grown, hygroscopic crystal. It grows slowly so that a great deal of time is usually required to obtain a good quality crystal of any size. For this same reason, crystals over 1 cm^2 are a rarity. The hygroscopic nature of the crystal requires additional care in the fabrication of finished pieces and the optical faces usually must be protected with cover slides which require fabrication and coating also. In addition to these problems, the crystals are fragile and care is necessary to prevent microcracks which will later grow in size during handling or pressure bonding processes. For these reasons, this material is usually reserved for use only where its higher figure of merit is necessary, or where its lower cut-off wavelength of approximately 3500 Å is necessary over $PbMoO_4$'s cut-off of approximately 4000 Å.

E. Lead Molybdate ($PmMoO_4$), n = 2.39, 2.27, V_s = 3.75 x 10^5 cm/sec., M_{2R} = 24

Lead molybdate[3,6] is fast becoming the most popular material for deflectors with large numbers of resolvable spots in the visible wavelength region. Its reasonably high figure of merit, low acoustic absorption (1 db/μsec @ 500 MHz), and availability in moderately large sizes are among its merits. It is a Czochralski grown crystal which may be grown in a single day, and its optical quality is continually being improved. Although somewhat fragile, it may be worked with reasonable care and good indium bonding transducer techniques. Its cut-off wavelength of approximately 3900-4000 Å is no problem for most of the visible lasers, and coloration is usually low so that optical powers of several watts may be tolerated. Present technology and crystals allow construction of 500 resolvable spot devices with access times of 6.8 μsec. The only major drawback of this material at this time is its cost and this, too, should improve with time and demand.

F. Tellurium, n = 4.8, V_s = 2.2 x 10^5 cm/sec., M_{2R} = 2920

Crystalline tellurium[7] has highest figure of merit for the 5-20 micron wavelength region. It is, however, somewhat like α-HIO_3 in that although it has the highest figure of merit, it has several disadvantages for many applications. One of the main disadvantages is the difficulty in getting moderately large pieces of good optical quality. In addition, tellurium has an optical absorption coefficient of 0.3 cm^{-1} at 10.6 microns--which makes it unsatisfactory for intracavity modulation or high power laser deflection at this wavelength. For these reasons, this material is usually reserved for those applications where a maximum figure of merit is required for use with a low power laser.

G. Germanium, n = 4.00, V_s = 5.50 x 10^5 cm/sec., M_{2R} = 540

Next to tellurium, crystalline germanium[8] is the best acousto-optic material for the 2-20 micron region. The figure of merit is 16 times greater than was originally thought--which accounts for its comparatively recent acousto-optic usage in comparison to its widespread use in lenses and windows for this wavelength.

In high power applications, it must be cooled in order to prevent thermal runaway. When this is done, however, an optical absorption coefficient of 0.032 cm^{-1} results. Further cooling can reduce this to as low as 0.01 cm^{-1}. An acoustic attenuation of 4.2 db/μsec. @ 500 MHz restricts the upper frequency, the bandwidth, and hence the large numbers of resolvable spots for short access times with this material. There is, however, a transverse mode of acoustic propagation which has an acoustic attenuation of 0.8 db/μsec @ 500 MHz (V_s = 3.51 x 10^5 cm/sec.), and which only reduces the M_{2R} to 190. The higher velocity of sound in germanium relative to tellurium decreases the difference between the two materials when alternate figures of merit M_1 or M_3 are computed. These figures have a different dependence on velocity of sound than M_2 when bandwidth or other considerations enter into the material selection. This material is readily available in large, good optical quality pieces, which make it an increasingly popular choice for this infrared wavelength.

H. Arsenic Trisulfide (As_2S_3), n = 2.38, V_s = 2.6 x 10^5 cm/sec., M_{2R} = 230

Arsenic trisulfide[9] is a special type of glass which finds its chief use in bridging the gap between visible and infrared materials. Although it has a relatively large figure of merit, it has an optical absorption coefficient of 0.7 cm^{-1} and an acoustic attenuation of 11 db/μsec. These shortcomings make this material generally suitable only for low power, and applications requiring low numbers of resolvable spots.

NEW MATERIALS

New materials are being explored for their use in acousto-optic devices. It is difficult to predict at this time which of these materials will be developed to a high degree of optical quality and low acoustic loss, as well as yield a high figure of merit for acousto-optic coupling. Highlights of some of these materials appear in Table I. Those interested in how one goes about looking for a new material are referred to the article by Pinnow[10] on this subject.

MISCELLANEOUS MATERIALS

There are many other materials which have been investigated which are too numerous to tabulate here. For those interested in other materials, Dixon[9] summarizes quite a number of them. Among those tabulated are GaAs, GaP, $LiNbO_3$, and TiO_2.

APPLICATIONS AND SYSTEMS CONSIDERATIONS

The applications of acousto-optic deflectors are many and varied. They include large screen laser-TV display, holographic memories in optical computers (ROM), computer output microfilm systems (COM), optical search and tracking radars, 3-D computer output displays, training device displays, and delay control in laser acoustic delay (LAD) lines. Although each system's application has its own requirements and freedoms, there are some aspects of acousto-optic deflectors, other than the N, τ, and η discussed earlier, which bear directly on the ease with which a given deflector, or choice of deflectors, may be incorporated into the system. Also, its accuracy in performing its function should be considered.

One systems consideration of a deflector is the angle which is scanned. As was indicated in the background section, the scan angle is not necessarily the same for two 500 spot deflectors of dissimilar materials, but rather only the $\tau(\Delta f)$ product is the same. In principle, the scan angle is not important since it may be magnified or demagnified by optical means. In practice, however, this magnification may become a design and cost factor since it governs the complexity and quality of the lenses which must be used. That is to say, larger magnifications may also require high quality, low f number lenses. Thus the angular scan may become a deflector selection factor in optimizing the system design.

A second consideration, closely related to the first, is the size and aspect ratio of the active aperture of the deflector. The reason that large aspect ratios arise is in order to conserve acoustic power and, hence, rf power and heating. As was discussed in the background section, the acoustic power required is inversely proportional to the height of the sound beam normal to the sound and optical beam propagation directions. Thus a 0.5 cm height will require only half the input power as a 1.0 cm height. On the long side of the aspect ratio, it is desirable to have a large aperture in the direction of sound propagation in order to provide the largest number of resolvable spots. These unequal dimensions require that the system optics provide diffraction limited cylindrical lenses for matching the normal gaussian laser beam to them and recollimating the beam afterward. A 2-dimensional scan increases the complexity even more. For these reasons and also due to the abundance of spherical optics relative to cylindrical optics, initial experiments are usually performed with near unity aspect ratio devices or, alternately, a somewhat higher aspect ratio device is uniformly illuminated and the additional optical loss is tolerated.

A third consideration is in the use of two acousto-optic deflectors to form an X-Y display. This consideration is the efficiency of a two stage deflector. The overall efficiency, whether two individual cells are used or one medium is used with two transducer sets, is the product (η_1, η_2) of the two single cells. The combination of two cells into one is only feasible in the case of unity aspect ratios since no intermediate optical elements are required. Even in this case, two cells are usually used for ease of alignment and "peaking" of the two characteristics. Also, with present fabrication techniques and the need for separate electronic drivers, very little cost saving is realized in combining the two into one.

Further considerations in acousto-optic deflectors are the distortions due to changes in the temperature of the device and thermal gradients across its aperture. Both are due to the variation of the sound velocity in the medium with temperature. Since the angle by which the optical beam is deflected is inversely proportional to the velocity of sound as described by the Bragg formula, a difference in the velocity of sound will lead to a difference in the deflection angle. In the case of a change in the temperature of the entire device, this results in a shift of the entire spot field and a small change in the scale factor associated with angular deflection of the beam per increment of frequency. Also changed by a small amount will be the access time and, hence, the number of resolvable spots. For the case of a thermal gradient across the aperture, all of the diffracted light will not be deflected in the same direction. In the case of lead molybdate, which has a temperature coefficient of longitudinal sound velocity of -161 ppm/^{0}C[13], a 10 degree gradient will cause spread of 0.8 spot in angle between deflected components from opposite sides of the aperture of a 500 spot deflector. Good thermal design of a deflector cell as well as conservation of sound power make for small thermal gradients. The latter distortion, due to thermal gradients, is the main concern in display type systems where resolution is the main concern but not necessarily the absolute location. In optical memory applications, both types of thermal distortion are of concern since location repeatability is also important. The overall topic of distortions, that is departures from the cw or ideal case, is just beginning to be analyzed and characterized. There is little doubt that as the use of these devices in systems increases, this characterization will rapidly develop.

SUMMARY

It has been the purpose of this article to bring together the necessary background material, a descriptive materials summary, and a discussion of the systems considerations which bear upon present day acousto-optic deflectors. To summarize the capabilities of a number of the materials discussed along with the corresponding parameters of interest from a systems standpoint, Table II has been prepared. Although the devices listed are not necessarily the largest devices which can be fabricated, they are typical of devices which are at the high end of resolvable spot capability for present day devices fabricated from these materials.

Present day deflectors provide larger, unmagnified, angular scans than electro-optic analog deflectors, have many unique features such as a random access capability, and have increasingly large efficiencies and numbers of resolvable spots available to the electro-optic systems designer. With these features and, in addition, the decreasing cost of these units as batch and mass production are initiated, the role of acousto-optic deflection seems destined to increase even more rapidly than in the past.

TABLE I.

Material	n	V_s (x10^5 cm/sec.)	M_2 (x$10^{-18} s^3/g$)	M_{2R}	Acoustic Absorption	Reference
TeO_2	2.3	0.617 (shear)	793	525	1 db/cm @ 25 MHz	11
TeO_2	2.3	4.26 (long.)	35	23	1 db/cm @ 30 MHz	11
$Sr_{.75}Ba_{.25}Nb_2O_6$	2.312, 2.299		39	26		12
Pb_2MoO_5			127	84		12
$Pb_5(GeO_4)(VO_4)_2$	2.281, 2.275	3.45	50.6	34	1 db/cm @ 200 MHz	12

TABLE II.

Material	Rayleigh Spots	Access Time (μsec)	Wavelength (Å)	Percent Deflection	Active Aperture (mm)	$\Delta\theta$(mr)
Quartz	1,000	20	6328	3-10	5 x 120	5.3
Water	200	12.5	6328	?	2 x 19	6.8
Glass	400	10	6328	60	2 x 38.6	6.5
α-HIO_3	100	2.1	4880	70	1 x 5	9.5
$PbMoO_4$	500	6.8	5145	50	4 x 26	10.1
$PbMoO_4$	100	1.35	4880	60	5 x 5	9.6
TeO_2	100	7.7	6328	13	5 x 5	14.1

REFERENCES

1. Korpel, A., et al., Proc. IEEE, 54, 1429 (1966).
2. Gordon, E. K., Proc. IEEE, pp. 1391, October 1966.
3. Heynau, H. A. and Barnard, G. M., Proc. EOSD Conf., 640, (1970).
4. Sittig, E. K. and Cook, H. D., Proc. IEEE, 56, 1375 (1968).
5. Pinnow, D. A., and Dixon, R. W., Appl. Phys. Letters, 13, 157 (1968).
6. Pinnow, D. A., et al., Appl. Phys. Letters, 15, 83 (1969).
7. Dixon, R. W. and Chester, A. N., Appl. Phys. Letters, 9, 190 (1966).
8. Abrams, R. L., and Pinnow, D. A., Jl. Appl. Phys., 41, 2765 (1970)
9. Dixon, R. W., J. Appl. Phys., 38, 5149 (1967).
10. Pinnow, D. A., IEEE Jl. Quantum Elec., QE 6, 223 (1970).
11. Uchida, N. and Ohmachi, Y., Jl. Appl. Phys., 40, 4692 (1969).
12. Watanabe, A., Nabeta, Y., and Yano, T., IEEE CLEA, Paper 5.9 (1971).
13. Sittig, E. K., Proc. EOSD Conf., 681 (1970).

CURRENT DESIGN CONSIDERATIONS FOR BULK ACOUSTO-OPTIC DEVICES IN DEFLECTION AND MODULATION APPLICATIONS

Dr. David E. Flinchbaugh
Control Laser-Orlando, Inc.
11222 Astronaut Boulevard
Orlando, Florida 32809

INTRODUCTION

Over the past forty years, and particularly since 1960, the recognition and utilization of the interaction of ultrasonic and optical beams has become increasingly important to the scientific community.[1] This significance stems from the fact that acousto-optic techniques permit amplitude, frequency, phase, polarization, and angular (spatial) modulation of laser beams. Any one, or combination, of these effects can thus be put to use performing vital functions in optical systems.[2,3]

The early research and development efforts have led into sophisticated engineering of laser modulation and beam steering devices designed for a variety of systems applications.[3] The basic versatility and dynamic electronic controllability of acousto-optic cells has established itself in such fields as visual computer displays, large screen television, radar and sonar signal processing, laser communications, data recording and printing, fluid velocimeters, laser machining of materials, and many other areas involving optical instrumentation and sources, including holography. This paper defines the design parameters employed in the current state of the art, starting with the optical diffraction cells employing very high frequency to microwave frequency ultrasonic waves, and mentioning also the refractive interaction which is useful in a smaller number of low frequency scanning or modulation applications.

BASIC PRINCIPLES

Let us consider first that we have a suitable material for bulk acoustic waves to travel in with minimal attenuation. In addition, it must be transparent to an optical beam traveling through it in a direction 90° to the ultra sound. The most useful range of ultrasonic frequencies for these devices is from a few megahertz to several gigahertz. The majority of current practical bulk wave units operate in the 20 to 200 MHz range.

These elasto-optical devices differ from the electro-optical types in that optical birefringence is induced by a propagating mechanical stress instead of an electric field creating polarization modulation or birefringence. Electro-optic deflection requires the use of devices such as Wollaston prisms to translate this modulation into two-position beam steering, and multiple prisms may be cascaded in binary arrays to make display systems.

On the other hand, ultrasonically generated phase diffraction gratings allow dynamic control of many beam positions by a single acousto-optic device operated over a moderate frequency bandwidth. The velocity of sound, which is a factor of approximately 10^5 slower than that of light, permits a large number of grating lines in moderately small apertures. These last two statements represent important considerations in device capacity for memory or display applications, as will be shown later. Another design factor relative to working aperture relates to instantaneous bandwidth, or response speed, which is a critical trade-off parameter when compared with electro-optic units for amplitude or frequency modulation applications.

The basic elements of an acousto-optic diffraction cell are illustrated in Fig. 1. A laser beam incident from the left has a width (or diameter) W along the upward traveling acoustic beam column which is much greater than the acoustic wavelength. The optical beam also has frequency f and wavelength λ in the medium.

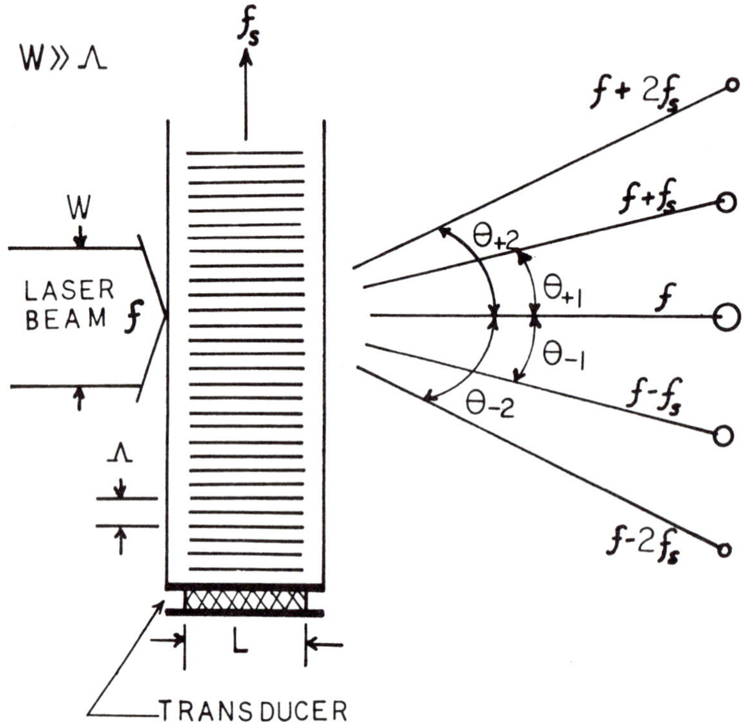

Fig. 1. Diffraction of light by an ultrasonic wave.

Because of its finite beam diameter W, there is a diffraction spread function $\delta\phi$ associated with the light which is important for two reasons: 1) it forms a basic criterion for the resolution of beam positions or number of spots, and 2) it plays a dominant role in the transition region when $W \approx \Lambda$, the acoustic wavelength. Note that the interaction at normal incidence (the Raman-Nath case) produces several Fraunhofer diffraction orders. The angular deflection relation is expressed as

$$\sin\theta = \pm\, m\frac{\lambda}{\Lambda} = m\left(\frac{\lambda}{V}\right) f_s \qquad (1)$$

where m = any integer defining the order number, and V = velocity of sound in the medium. The exiting light is both phase and frequency modulated as given by

$$E' = E_o \sin\left\{2\pi ft + \frac{2\pi L}{\lambda}\left(n_o + \Delta n \sin 2\pi f_s t\right)\right\} \qquad (2)$$

where n_o is the refractive index of the medium at equilibrium, L is the distance the light travels in the acoustic beam, t = time, and Δn is the change in refractive index produced by the acoustic stress wave. The diffracted beam intensity for each order is given by the Bessel function

$$I_m = J_m^2\left\{\frac{2\pi L(\Delta n)}{\lambda}\right\} \qquad (3)$$

For most applications it is wasteful to allow the optical beam to be scattered into many orders. If one adjusts the beam interaction angle just off of normal incidence to the point where the emergent beam appears at twice the incident angle, then

$$\sin\theta_B = \frac{\lambda}{2\Lambda} = \left(\frac{\lambda}{2V}\right) f_s \qquad (4)$$

This is called the Bragg angle, and this condition is analogous to the law of reflection since the incident and emergent beams are at equal angles to the normal, or acoustic beam axis. Provided that L is sufficiently large, constructive interference is observed and the Bragg angle favors diffraction into a single order. An optical quality factor has been defined[4] for this effect, as

$$Q = \frac{\lambda L}{n_o \Lambda^2} \qquad (5)$$

Perhaps a more useful expression is the following[5], where λ_o is the free space optical wavelength and Δf is the usable acoustic bandwidth.

$$L = \frac{n_o V^2}{\lambda_o f_s \Delta f} \qquad (6)$$

Using this criterion, the sound aperture L gives an adequately wide diffraction spread and there is only about a 1 dB drop in the interaction level at the band edges.

It is important to note from Equation (1) that the angle of deflection depends on the ultrasonic frequency, and thus the angular range of control is a function of the operating bandwidth. It is worthwhile to consider the resolution capability of such a deflector. Making the small angle approximation, which is valid until one reaches microwave acoustic frequencies,

$$\delta\theta = 2\Delta\theta_B = \frac{\lambda\Delta f}{V} \qquad (7)$$

Then we concern ourselves with the diffraction spread mentioned earlier and define this factor in terms of the beam size, shape, and wavelength according to the Rayleigh criterion for resolution

$$\delta\phi = K\frac{\lambda}{W} \qquad (8)$$

For a circular aperture, K = 1.22 but using the approximation for an extended restangular aperture, K becomes unity. Thus we simply divide the minimum beam width into the maximum angular range to determine the number of resolvable spots N:

$$N = \frac{\delta\theta}{\delta\phi} = \Delta f\frac{W}{V} = (\Delta f)\tau \qquad (9)$$

The time τ represents the transit time for the acoustic wave, moving at velocity V, to cross the optical beam aperture W. A device having a transducer and associated components capable of performing over 80 MHz bandwidth with an aperture transit time of 10 microseconds can project up to 800 resolved spots in a display system or will store an equal number of bits in an optical memory system.

PRACTICAL DEVICE SPECIFICATIONS

The primary performance objectives are simply stated: (a) high efficiency of optical modulation, and (b) large bandwidth of operation. Other important considerations include: 1) low electrical drive power requirements, 2) minimum optical aperture size consistent with meeting resolution requirements, 3) low optical wavefront distortion over the entire active aperture during both static and dynamic operating conditions, 4) minimal passive optical losses, and 5) compact size for system compatibility.

Operating bandwidth is somewhat more limited for the acousto-optic case than for electro-optics, which can be base-band devices up to some high frequency cutoff due to capacitance, high voltage switching, or some other driver or crystal limitation. The choice of a center frequency, such as 120 MHz allows one to design for a usable bandwidth of about one octave, or ± 40 MHz. This is a reasonable figure when one considers the undesirable possibility of having overlapping diffraction orders generated by ultrasonic harmonic frequencies if the range was extended beyond the 80-160 MHz example. Also, it requires a great deal of skill just to achieve a one octave frequency range in handling the mechanical impedance matching of the transducer/interaction medium, the electrical drive impedance coupling, and the optical Q matching to within a 3 dB response characteristic.

Let us consider the acousto-optical interaction materials which are used currently. Table I lists the most widely employed media for this application and makes a comparison of their relative efficiencies.

Material	Relative Acousto-Optic Figure of Merit	Compressional Wave Acoustic Velocity
Fused Silica (vitreous quartz)	1.0	5.95×10^5 cm/sec
Sapphire (c-axis sound)	0.2	11.00×10^5 cm/sec
Lithium Niobate	4.6	6.57×10^5 cm/sec
Dense Flint Glass	6.4	3.80×10^5 cm/sec
Lead Molybdate	23.7	3.75×10^5 cm/sec
Distilled Water	106.0	1.5×10^5 cm/sec

Table I. Selected Materials for Acousto-Optic Devices

The figure of merit M for a given deflector medium is defined in terms of the refractive index, acoustic velocity, the photoelastic constant p, and the density ρ.

$$M = \frac{n_o^6 p^2}{\rho V^3} \tag{10}$$

These numbers are normalized ($\bar{M}$) to fused silica, which has long been a standard of the industry, for comparison purposes. Optical quality fused silica is selected for applications requiring a large aperture, eg. 5 cm or more, at frequencies from 10-120 MHz. Single crystal sapphire is used for the microwave acoustic units up to nearly 10 GHz since the attenuation of ultra sound is very low along crystalline axes. Lithium niobate is a good high frequency material but is relatively costly and subject to limitations on incident laser power due to an observed damage threshold. Due to their ultrasonic attenuation properties, the glasses are limited to around 50 MHz, depending on aperture size. It is undesirable to have a large amplitude gradient across the working aperture. Lead molybdate single crystal has received the most attention in the past year.[6] Acoustic loss at 500 MHz is only 1.0 dB/cm, which makes it ideal for use in VHF-UHF systems. A number of acousto-optic devices having deflection efficiencies greater than 50% over an 80 MHz bandwidth with less than 1 Watt of electrical drive power have been made for use with argon ion (green) and other laser systems. We notice that water, the only liquid on the table, has the highest figure of merit of all due to its high compressibility. This property is exhibited by most liquids compared to solids. However, the primary drawback is that ultra high frequencies are attenuated severely in liquid media. Also, convection currents tend to distort the optical wavefront as acoustic drive levels increase.

It is desirable to have a relation for calculating the acoustic power P_a (in watts) and diffraction efficiency η in terms of the geometrical constants and the materials properties. Expressions for these quantities have been developed in Ref. 7 and have subsequently been abbreviated to the following form for convenience in illustrating the conditions for 70% deflection of the incident light beam for the Bragg case.

$$P_a = \frac{54H}{LM}\left(\frac{\lambda}{0.6328}\right)^2 \qquad (11)$$

where λ is expressed in microns and H is the aperture beam dimension normal to W. Then

$$\eta = \sin^2\left(\frac{P}{P_a}\right)^{\frac{1}{2}} \qquad (12)$$

For 100% deflection $P/P_a \approx 2.5$, but it is evident that as L increases the response will drop off at the band edges. Thus there are design trade-offs and practical limitations in transducer drive level, thermal rise and gradient tolerances before bonds disintegrate or other material damage occurs, and optical distortion tolerances. Because of the dependence on λ, more power is required to make an efficient device at longer wavelengths. It is often necessary to use forced fluid cooling of the deflector/transducer assembly in addition to careful thermal coupling of the solid state elements. Most of these devices are traveling wave types where the ultrasonic energy must be absorbed in a terminating block after traversing the optical interaction volume. Thus heat must be dissipated from there as well as from the partial conversion loss at the transducer end.

Note that at the bottom of Fig. 1 a simple but very useful type of transducer is illustrated. It takes the form of a plane wave-generating transducer consisting of a flat plate of piezoelectric material sandwiched between two electrodes. When an electric field is applied in one direction across this material, it will expand; when it is applied in the opposite direction (or polarity), the material will contract in that same dimension. These dimensional changes are coupled into the transparent medium above and travel through the medium compressional (or longitudinal) mode acoustic stress waves. One can also design and use transducers that generate shear waves. For that case, the particle motion in both the transducer and the medium, instead of being along the direction of ultrasonic wave propagation, is transverse (or 90°) to the beam path. The longitudinal wave case is easier to visualize. However, there are some advantages in using shear waves since one can employ the rotation of the optical beam plane of polarization as the interaction occurs. One can infer from this diagram that there are both mechanical coupling considerations from the transducer to the medium and electrical coupling considerations in connecting the electrodes of this variable parallel-plate capacitor device to an external circuit.

The choice of a suitable electro-acoustic transducer material is based primarily on the properties listed in Table II. The piezoelectric coupling coefficient gives a measure of the efficiency of converting an electrical signal into an acoustic wave, the dielectric constant is important from an electrical circuit element and impedance matching standpoint, and the frequency thickness constant indicates what dimension the material must be fabricated to in order to operate in the desired frequency range. The mechanical impedance, defined by the product (ρV) is also significant because it will present a boundary layer to the coupling of acoustic waves into the deflector when there is a mismatch. The most widely used material has been crystalline quartz of various cuts, a Y-cut producing a transverse sound wave and an X-cut producing primarily a compression sound wave. A class of materials prepared in the form of powered sintered ceramics that can be molded into various shapes and sizes, and then electrically poled, include potassium zirconate titanate (PZT), potassium sodium niobate (PSN), and barium titanate. These transducer materials have relatively high piezoelectric coupling coefficients and are used extensively. They suffer, however, from a very high dielectric constant which makes them behave as high value capacitors in thin sections. Also, they are not as mechanically cohesive as single crystals when lapped to very thin dimensions, and thus there are definite fabrication limitations in the 100-150 MHz range. Their Curie temperatures, above which they no longer behave piezoelectrically, are in the order of 120°C which means that either low drive levels or well-engineered cooling must be employed. Lithium niobate, used as a wafer cut from a single crystal, is now widely utilized for both shear and compressional wave transducers, depending on the crystalline orientation. It is currently the best choise for lead molybdate and other high-performance devices. The last two on the list are cadmium sulfide and zinc oxide. These materials can also be used as fabricated single crystals, or they can be vapor deposited onto a substrate in a high vacuum deposition chamber. These thin films then can be used to generate very high frequencies in the

Material	Coupling Coefficient	Relative Dielectric Constant	Frequency Thickness Constant in MHz-mm
Crystalline Quartz Y-cut (shear)	0.137	4.58	1.87
$LiNbO_3$ Single Crystal (shear)	0.54	80	2.08
Potassium-Zirconate-Titanate - 7 (shear)	0.68	840	1.00
PZT-5 (compressional)	0.75	1700	1.77
Cadmium Sulfide (compressional)	0.154	9.53	2.2
Zinc Oxide (compressional)	0.34	8.84	3.2

Table II. Piezoelectric Transducer Materials for Acousto-Optic Devices

microwave region (up to around 10 GHz or X-band) and are often used on sapphire substrates.

Transducer design considerations call for the material thickness to be one-half of the acoustic wavelength in that medium for the fundamental mode of operation. Thus, for example, lithium niobate is about 1.5×10^{-3} inch for a transducer operating in compressional mode at a frequency of 100 MHz. The velocity of propagation of ultra sound in these crystalline media is generally slower for shear waves than for longitudinal waves. One can see that for microwave applications, thicknesses in the micron region are required for fundamental wave transducers; thus, thin films must be employed. Plane, parallel plate transducers operating at high frequencies are often segmented (normal to the optical path) into an array of series elements to reduce the capacitive reactance of the circuit.

For transducers which are fabricated in the VHF range, they are generally handled as oriented wafers down to thicknesses of about 10^{-2} inch. After a fine lapping on one side they are bonded to a precision lapped substrate surface (the interaction medium). The finish tolerances are kept to a small fraction of Λ in order to minimize acoustic wavefront distortion. Certain epoxy compounds which form a very strong, thin glue line have proven to be satisfactory for this purpose. An ideal coupling medium, however, has an acoustic impedance value equal to the geometric mean of the transducer and substrate impedances. This usually calls for a metal such as tin, lead, indium, or special alloys. Intermediate layers which are analogous to quarter-wave (for broad banding the response characteristic) or half-wave tuning in electrical circuits are devised to optimize acoustic energy coupling and bandwidth for specific applications. Indium cold welding in high vacuum under high pressure produces very thin, or controlled thickness, bonds of excellent quality. We have, along with other laboratories,[5,6] found this to be the most satisfactory technique for accomplishing the most critical bonding operations in state-of-the-art devices. Additional details relative to the mechanical and electrical coupling parameters may be found in Ref. 12.

Table III illustrates a set of typical specifications for an up-to-date, high performance, acousto-optic diffraction cell available commercially from our laboratories. This type of unit may be used either as a deflector or modulator, within the limits of its geometric and operating parameters, and those of the optical system which it becomes a part of. As a beam deflector, the full aperture must be utilized to achieve 100 spot resolution. However, one can realize a much smaller transit time, and hence larger instantaneous bandwidth response, by focusing and recollimating the optical beam to a very small diameter for modulation purposes. In this manner, amplitude or frequency modulation is effectively done at high frequencies with low voltage signals, thus eliminating some of the problems associated with electro-optic techniques. Numerous applications are discussed in the references cited,[8,10,11,12] and some details of the associated optics and electronics are illustrated schematically.

Material: lead molybdate single crystal, lithium niobate transducer
Bragg angle θ_B for λ = 5145 Å (Argon green): 0.6 degree
Size overall: 3.2 cm x 3.5 cm x 7.6 cm

Center frequency f_s	130 MHz	Number of resolvable spots	100
Bandwidth Δf	80 MHz	Maximum scanning rate	100 KHz
Aperture W	5 mm	Passive optical loss	<1%
Transit time τ	1.35 μs	Maximum cw rf drive	1.5 watts
Random access time	1.35 μs	Radiation cooled package	
Angular scan $\Delta\theta$	10 mrad	Deflection efficiency	$\geq$50%

Table III. Typical Specifications for a Current Acousto-Optic Diffraction Cell

SPECIAL DISPLAY TECHNIQUES

In order to produce the greatest deflection efficiency uniformly over the largest practical bandwidth at minimal power levels, two schemes for steering the acoustic beam to track the Bragg angle variation with frequency have been explored.[7,8,9] Phased array transducers are employed as depicted in Fig. 2. Here the transducer segments are equally spaced by a distance $S = \Lambda_0^2/\lambda$. This scheme combines and simplifies the earlier techniques[9] of separate phase shifts between driven elements and staircase blazing of the array. The composite beam steering array illustrated is composed of two 5-element subarrays, each of which produces 1st diffraction order acoustic beam steering with stepped transducers. The phase shift Ψ is adjusted as a function of frequency to follow the envelope of the response curve. In this manner the 3 dB bandwidth is extended to 105%. At any given time the ultrasonic wavefronts generated by the array form a staircase, which in the far field is equivalent to a planar wavefront making an angle α with the transducer faces. The blazed transducer structure, unfortunately, is costly and difficult to fabricate.

Figure 3 illustrates an ultrasonic cell designed for sinusoidal beam scanning by dynamic optical refraction.[10] This case requires that

$$W \ll \Lambda \qquad (13)$$

The laser beam responds to moving regions of alternate compression and rarefaction, and this buld refractive index perturbation results in a deflection through the angle $\pm\theta_R$.

$$\theta_R = \frac{2\pi K'L}{\Lambda} \frac{(n_o^2 - 1)(n_o^2 + 2)}{6n_o} \Delta P \qquad (14)$$

In this expression, K' = compressibility constant of the medium and ΔP = variation in pressure produced by the ultrasonic stress wave. Liquids are the most suitable media for this technique due to their large values of K'. The most useful frequency range is 100 KHz to 1 MHz. Note that the deflection angle is a function of drive level, or sonic amplitude, and that rapid sequential scanning is possible. The beam stops and changes direction at the maximum end point angles, and the scan rate is relatively linear in the central 35% of deflection around the equilibrium beam position. This type of device is inexpensive and simple to build, however, the spot resolution is limited to about 100 and the deflection to $\pm 1^{0}$ without significant beam distortion.

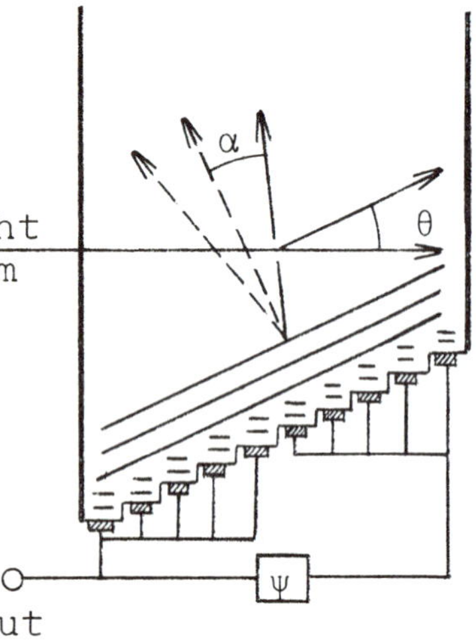

Fig. 2. Composite phased array transducer with blazing for sonic beam steering.

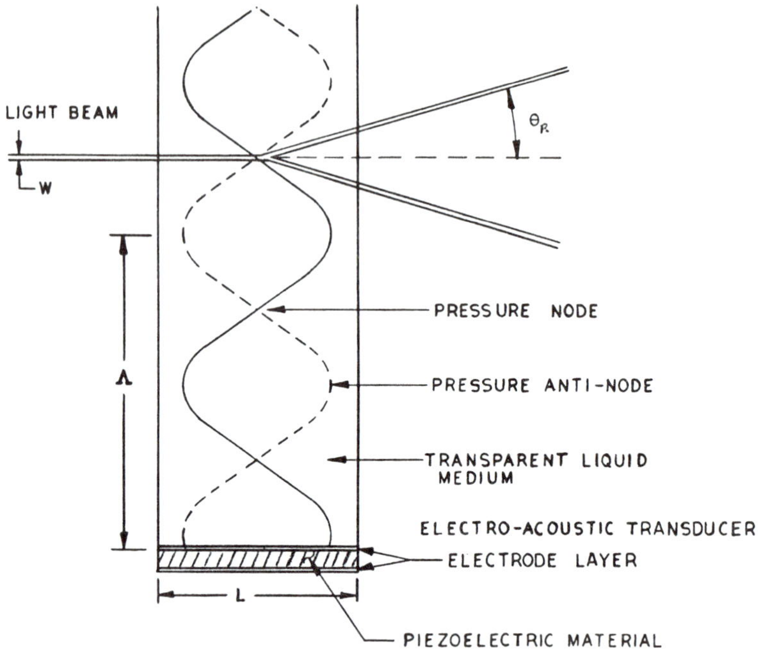

Fig. 3. Ultrasonic deflection of a light beam by refraction.

SUMMARY

At the present time we can conclude that existing bulk solid state acousto-optic devices achieve 100 spot resolution with an aperture of just over 1 microsecond, which can be scaled up to N = 500 with existing lead molybdate crystals. Fused silica modulators with N = 1000 and N = 4000 spots have also been demonstrated. Sapphire devices of several hundred spot capability are also available in the 1 GHz range. Response times in the order of 10 nanoseconds have been demonstrated at very small apertures. Existing deflectors are adequate to accomplish modulation and line scan functions in large screen laser TV displays.

Problem areas center around unwanted heat dissipation which degrades continuous wave optical performance, increased ultrasonic absorption in materials at higher frequencies, and relatively high materials and fabrication costs. However, current research and development may double the bandwidth and capacity with existing materials, and there is no doubt that the serious application of manufacturing engineering efforts to production techniques will ultimately bring electronically controlled optical beam modulator cost to a level compatible with that of rotating and mechanical types.

BIBLIOGRAPHY

1. Raman, C. V. and Nath, N.S.N., "The Diffraction of Light by Sound Waves of High Frequency, Pt. II", Proc. Ind. Acad. Sci., Vol. 2A, p. 413 (1935).
2. Danielson, G. E., DeMaria, A. J., and Flinchbaugh, D. E., "Use of Focused Acoustic Waves for Internal Control of High Gain Lasers", Proc. Intl. Symp. on Laser Physics & Applications, Berne, Switzerland (1964).
3. Flinchbaugh, D. E., "Acousto-Optic Laser Modulation", Laser Focus, Vol. 3, pp. 25-29 (Sept. 1967).
4. Klein, W. R. and Cook, B. P., "A Unified Approach to Ultrasonic Light Diffraction", IEEE Transactions, Vol. SU-14, p. 123-134 (July 1967).
5. Sittig, E. K., "Design, Technology, and Limitations of Elasto-optic Light Deflectors", Proc. Electro-Optical Systems Design Conf., NYC (Sept. 1970).
6. Pinnow, D. A., Van Uitert, L. G., Warner, A. W., and Bonner, W. A., "Lead Molybdate: A Melt-Grown Crystal with a High Figure of Merit for Acousto-Optic Device Applications", Appl. Phys. Let., Vol. 15, pp. 83-86 (Aug. 1969).
7. Gordon, E. I., "A Review of Acousto-optical Deflection and Modulation Devices", Proc. IEEE, Vol. 54, p. 1391 (1966).
8. Korpel, A., Adler, R., Desmares, P., and Watson, W., "A Television Display Using Acoustic Deflection and Modulation of Coherent Light", Proc. IEEE, Vol. 54, pp. 1429-37 (1966).
9. Coquin, G. A., Griffin, J. P., and Anderson, L. K., "Wide Band Acousto-optic Deflectors Using Acoustic Beam Steering", IEEE Transactions, Vol. SU-17, pp. 34-40 (1970).
10. Flinchbaugh, D. E., "Focusing Ultrasonic System Applicable to Two-Dimensional Optical Beam Scanning and Laser Output Modulation", J. Acoust. Soc. Am., Vol. 37, pp. 975-85 (June 1965).
11. La Macchia, J. T., and Coquin, G. A., "Simultaneous X, Y Acousto-optic Deflection", Proc. IEEE (Letters), Vol. 59, p. 304 (1971).
12. Flinchbaugh, D. E., "Laser Beam Control Utilizing Optical Diffraction and Refraction by Ultrasonic Waves", Proc. Intl. EOSD Conf., pp. 133-144 (March 1971).

PRACTICAL CONSIDERATIONS IN SPECIFYING AN ACOUSTO-OPTIC DEFLECTOR

A. Korpel, R. L. Whitman and M. Odom
Research Department
Zenith Radio Corporation
Chicago, Illinois

Introduction

Resolution capabilities of optical scanning systems are dependent on the intensity distribution of the light source. Deflection system resolution specifications generally assume an evenly illuminated input aperture. Since this is seldom the case in the ultimate system application, it is of interest to know how to modify such specifications for practical cases. This article discusses a straightforward technique for obtaining a meaningful resolution definition in terms of either a number of "resolvable spots" or a system modulation transfer function (MTF). Resolution measurement techniques and experimental data on a "130 spot" light deflector are presented. The data confirm the results predicted by the outlined approach.

While deflection resolution requirements dictate that the light beam be as wide an possible in the plane of deflection, deflection efficiency requirements indicate that the light beam be narrow in the orthogonal direction. The theoretical reasons for this resultant optical "sheet beam" requirement are discussed, and two optical systems which were used for generating such a beam are described.

The Bragg angle tolerance requirements which limit the acoustic bandwidth are outlined and discussed in connection with an experimental deflector.

A short section is devoted to access time considerations when a random access deflection system is desired.

Summary of Theory for the Evenly Illuminated Deflector

The number of resolvable deflection angles, i.e. the resolution N of an evenly illuminated deflector, is given[1] by the ratio of the total deflection angle, $\theta_d = (\lambda/V_s)\Delta f$, to the angular spread inherent in the light beam, $\theta_s = \lambda/A$. (Here λ is the wavelength of the light, A is the deflector aperture width and Δf is the total possible change in sound frequency.) Thus

$$N = \frac{\theta_d}{\theta_s} = \Delta f \frac{A}{V_s} \qquad (1)$$

It can be seen that the resolution, N, goes up as the aperture or light beam width, A, increases. The light beam need only be wide in the direction of sound travel so that its angular spread in the plane of deflection is small. The beam dimension orthogonal to this should be small enough to fall completely within the acoustical field as determined by the transducer height. For reasons of deflection efficiency (as will be discussed later) the transducer height should be small. This results in the "sheet beam" geometry for the light shown in Fig. 1.

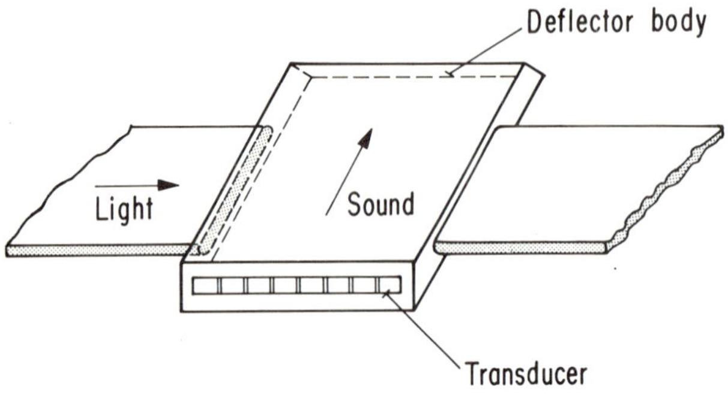

Fig. 1. Acousto-optic Light Deflector

Rayleigh and 40% Criteria for Resolution

The value for N was obtained from (1) by assuming that the deflector aperture of width A is evenly illuminated by the light. The far field intensity pattern $|E(x)|^2$ of such an aperture distribution, observed for instance in the focal plane of a lens with focal length F, is of the form

$$|E(x)|^2 \propto \left[\frac{\sin \frac{\pi A x}{F\lambda}}{\frac{\pi A x}{F\lambda}} \right]^2 \tag{2}$$

as shown in Fig. 2.

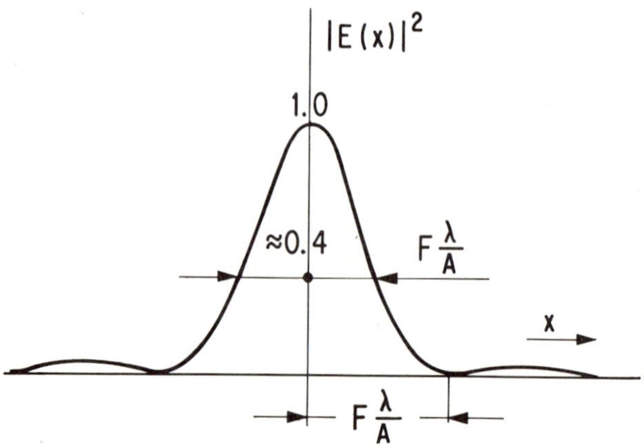

Fig. 2. Far Field Intensity Pattern of Evenly Illuminated Aperture of Width A

The separation between resolvable spots is defined as that one which places the peak of one intensity distribution on the first zero of the other. This is the Rayleigh criterion[2] for resolution and results in the two resolved intensity distributions overlapping at the points at which their intensities are down to approximately 40% of that at their peaks, as shown in Fig. 3.

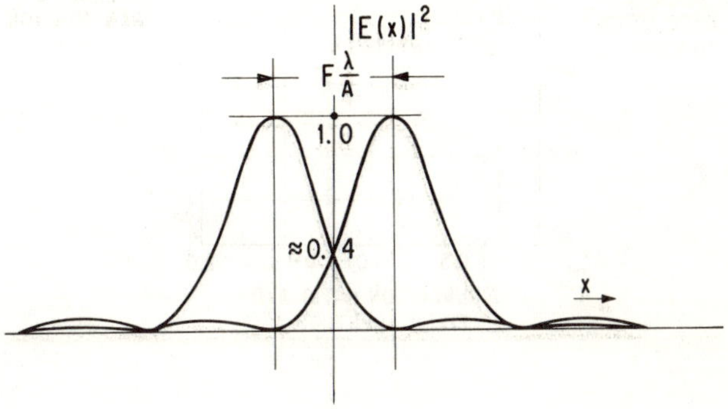

Fig. 3. Two Far Field Patterns, Resolvable by Rayleigh Criterion

Taking the spot width at the 40% intensity points as a measure of resolution leads to a criterion identical to the Rayleigh criterion where the latter may be applied (i.e. with a $(\sin x/x)^2$ distribution). However, the 40% criterion has the advantage of being applicable to more general distributions. It is also relatively easy to determine experimentally.

In most applications the finite aperture of the deflector, rather than being illuminated uniformly, is illuminated by a laser beam having a Gaussian intensity distribution (truncated Gaussian). This results in a broadening of the far field distribution, for instance as measured by the width between the 40% points. The factor which expresses this broadening depends on the truncation ratio $\rho = A/D$ where D is the diameter of the laser beam measured at its $1/e^2$ intensity point. The inverse of this factor, i.e. the resolution relative to the evenly illuminated aperture case, is plotted in Fig. 4 as a function of ρ[3,4]. In the same figure is plotted the relative optical transmission resulting from the truncation of the one-dimensional Gaussian beam.

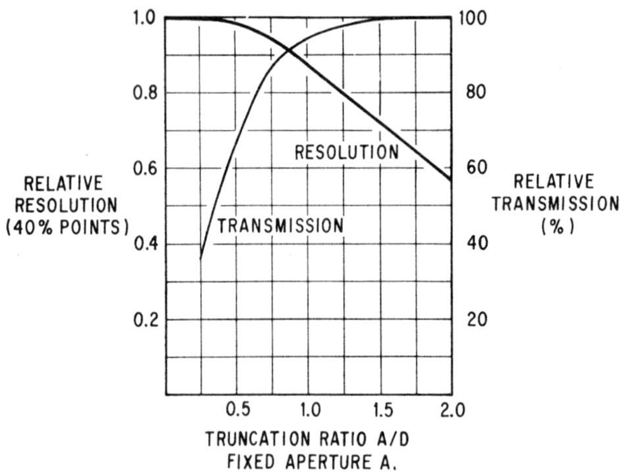

Fig. 4. Deflector Resolution and Transmission as a Function of Truncation Ratio

Modulation Transfer Function

The 40% intensity criterion for resolvable spot width was used above because it is relatively easy to determine experimentally. It should be noted however, that the use of just one parameter (such as number of resolvable points as defined here) gives only an approximate indication of the system's performance and, in fact, may be misleading in certain applications. A more complete measure of an optical system's resolution capabilities is given by its modulation transfer function(5) or MTF. One way to describe the MTF of a light deflector is to first consider an application to optical display of electrical information. Specifically assume the intensity I(t) of the light entering the deflector to be varying with time such that

$$I(t) = I_o (1 + a \sin 2\pi f t), \text{ with } a < 1 \qquad (3)$$

The time averaged intensity pattern seen in the focal plane of a lens following the scanner will then be

$$I(x) = I'_o [1 + a'(\varkappa) \sin 2\pi \varkappa x] \qquad (4)$$

where the symbol $\varkappa$ represents a spatial frequency (cycles/unit length) and x is the coordinate along the scan direction. The ratio $a'(\varkappa)/a$ is defined as the system MTF, and may be thought of as the optical spatial frequency response of the scanner when used as a component in any system for producing some desired intensity distribution. The spatial frequency $\varkappa$ is related to the temporal frequency f by

$$\varkappa = \frac{f}{V_{sc}} \qquad (5)$$

where V_{sc} is the velocity of the scanning spot in the focal plane. The MTF can be shown to be related to the autocorrelation function of the light amplitude distribution in the scanner aperture, in the back focal plane of the lens[5]:

$$\mathrm{MTF}\ (\varkappa) = \frac{a'(\varkappa)}{a} = \frac{\int_{-\infty}^{+\infty} E\ (u)\ E^*\ (u - F\ \lambda\varkappa)\ du}{\int_{-\infty}^{+\infty} E\ (u)\ E^*\ (u)\ du} \tag{6}$$

In this relation E is the optical field, u is the coordinate in the aperture plane, F the focal length of the lens and λ the wavelength of light. In case the aperture has a finite physical width A, the MTF always drops to zero at a limiting spatial frequency $\varkappa_c$ given by

$$\varkappa_c = A/F\lambda \tag{7}$$

Note that this number is the inverse of the resolvable spot size as determined by the Rayleigh criterion; also that $\varkappa_c$ in no way depends on the actual distribution E (u) as long as the extent of the optical field is physically limited by A only. Thus the limiting spatial frequency expressed in cycles per line scan is seen to equal the number of resolvable points (Rayleigh criterion) for an aperture A. The particular form of the MTF within this limited spatial frequency range depends, of course, on the character of the illumination E(u). Figure 5 shows the MTF for a Gaussian beam for various values of truncation ratio ρ = A/D. The curve is plotted as a function of the normalized spatial frequency $\varkappa/\varkappa_c$.

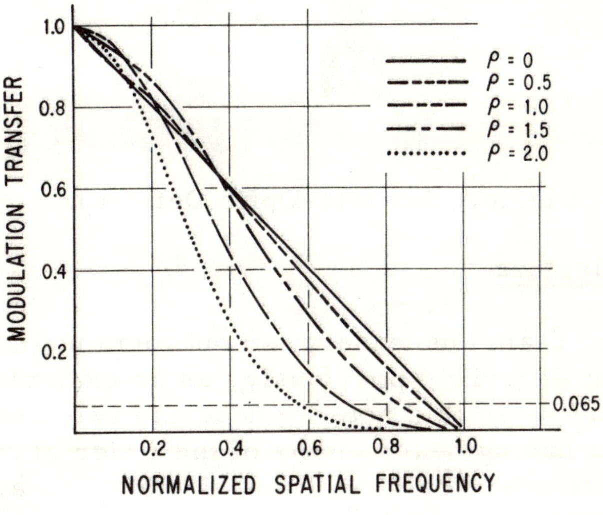

Fig. 5. Deflector Modulation Transfer Function for Several Truncation Ratios

As an example consider the deflector shown in Fig. 6, which was used in the experiment. This had an aperture width A = 2.5 cm, a bandwidth Δf = 20MHz and a sound velocity V_s = 3850 m/sec. Then from (1) N $\cong$ 130. So for this deflector the limiting spatial frequency in cycles/line scan equals 130. Using this number and the graph of Fig. 5 enables one to assess the relative performance at any spatial frequency for a range of truncations.

Relation Between MTF and 40% Criterion

It was already stated above (in conjunction with eq. 7) that the limiting spatial frequency (MTF = 0) expressed in cycles per line scan is equal to the number of resolvable points by the Rayleigh criterion. Between the more general 40% criterion and the MTF there exists a rough quantitative relationship. It may be shown that the number of cycles per line scan at the spatial frequency where the MTF response is down to 6.5% corresponds very closely to the number of resolvable points determined with the 40% criterion regardless of the truncation ratio. This may be verified readily from graphs 4 and 5.

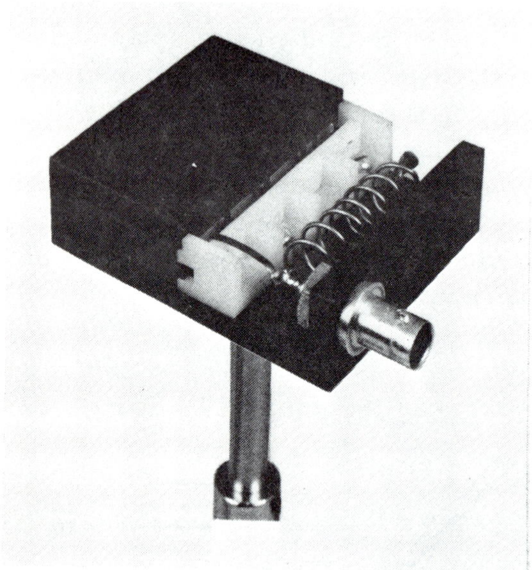

Fig. 6. Acousto-Optic Deflector

Optics for Light Deflectors

The optical sheet beam, necessary for optimum utilization of a typical scanner, such as that described previously, as an example may be generated in a number of ways. In this note we describe two methods, useful for operation with lasers having beam widths of the order of millimeters. In both methods the system is, for simplicity, designed so as to make the diameter of the scanning output beam equal to that of the input beam. The first configuration is shown in Fig. 7.

The light beam from the laser is first magnified by a spherical telescope focused at infinity (lenses L_1 and L_2). The resultant large collimated beam is transformed to a sheet beam in the focal plane of the cylindrical lens L_3. Upon passing through the scanner, the beam is recollimated by cylindrical lens L_4 located one focal length away from the scanner. The collimated beam at lens L_4 is then passed through a telescope to decrease the size of the output beam to that of the input beam. This automatically increases the scan angle.

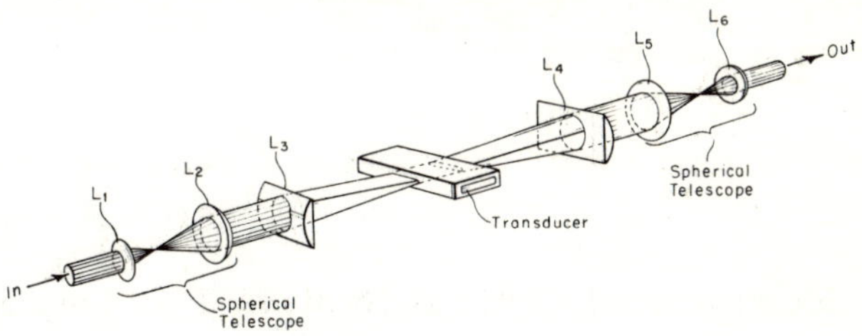

Fig. 7. Optical System For Deflector

Aberration caused by any of the lenses will broaden the scanning spot and degrade the system resolution. Although corrected spherical lenses, which will yield a diffraction limited spot, are readily available and relatively inexpensive, the same is not true for cylindrical optics. The most economical approach here is to choose a lens which may be used at a high enough f number (ratio of focal length to aperture diameter) to be free of aberrations. A rough guide to the minimum f number ($f_{\#}$) allowable for a plano convex lens under conditions of spherical aberration only, is given by the following relation:

$$f_{\#} > 0.30 \left[\frac{D}{\lambda} \right]^{1/3} \tag{8}$$

where D is the diameter of the laser beam incident upon the lens (the $1/e^2$ width for a Gaussian beam). In practice, for operation with a 25 mm aperture, it should be sufficient to use a simple uncorrected f/10 cylinder lens. Note that in the system of Fig. 7, the aberration broadening is predominantly in the direction perpendicular to the scan. Hence, unless a deflector is used as a component in a raster scanning system, relation (8) may possibly be relaxed considerably. The angular width of the focused sheet beam must not, however, be so large that part of the light misses the sound field. This condition sets a lower limit of ≈f/9 with the deflector of Fig. 6 in the configuration of Fig. 7.

A simpler and cheaper system but one more difficult to align and more susceptible to aberrations, is shown in Fig. 8.

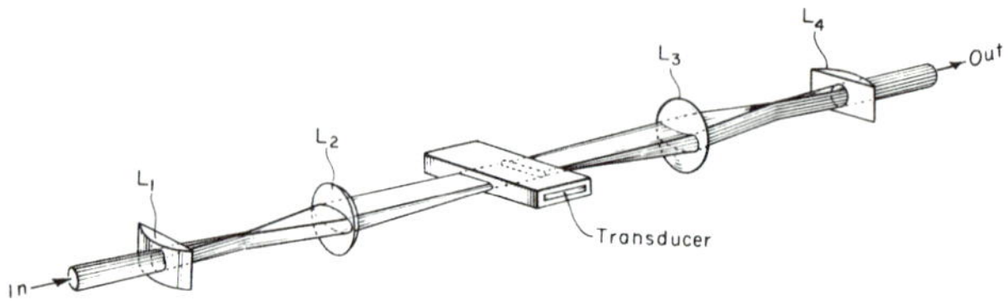

Fig. 8. Economical Optical System For Deflector

The required broadening of the light beam is here achieved by a small cylinder lens L_1 at the laser exit. After spreading, the beam is recollimated in the direction of sound propagation by spherical lens L_2 placed confocally with L_1. In the direction perpendicular to the sound sheet beam the light is focused at the center of the sound cell. An identical system L_3 and L_4 reshapes the beam and increases the scan angle. Note that aberration broadening in this system is in the direction of scanning; also that coma in uncorrected lens L_4 is of importance as the output beam is deflected over a considerable angle ($\approx 125 \times 10^{-3}$ rad for a one millimeter beam).

Experimental Confirmation

Because of its inherent simplicity, the system of Fig. 8 is very attractive. For this reason it was set up in the laboratory for resolution measurements on a scanner such as described in our previous example. The input laser beam originated from a Spectra-Physics laser (model 123) and had a Gaussian beam diameter of 0.73 mm as measured by its diffraction spread. Cylinder lens L_1 had a focal length of 8.5 mm and spherical lens L_2 one of 17.2 cm. The resulting sheet beam was 14.8 mm wide ($1/e^2$ points) at the scanner and truncated by the latter (25 mm) with a truncation ratio of 1.69. A slit was mounted on a calibrated translation stage and placed in the focal plane of a 351 mm focal length lens following L_4. The d.c. current out of a photodiode following the slit, being proportional to the light intensity at the slit, gave an indication of when the slit was at the peak and the 40% intensity positions of a deflected spot at any given scanner drive frequency. The 40% intensity widths taken at 31 MHz, 40 MHz and 49 MHz were .284 mm, .252 mm and .274 mm, respectively. The total "scan width" between the extreme frequency points, 49 and 31 MHz, was 1.94 cm. As may be seen, lens aberrations start to broaden the spot at 31 and 49 MHz and, in fact, become objectionable at 30 and 50 MHz. Based on the above measurements the number of resolvable points per line scan (40% criterion) lies between 66 and 74. For the 18 MHz frequency sweep used here the number of resolved points for a uniformly illuminated aperture may be calculated from (1) to be ≈ 117. For a truncation ratio of 1.69

the relative resolution as read off Fig. 4 is ≈0.65. Hence the number of resolvable spots (40% criterion) should be ~0.65 x 117 = 76. This is in good agreement with the experimental results. On the basis of this experiment it may be confidently expected that even with the simple configuration of Fig. 8, higher resolution, approaching the theoretical limit of 130 spots, may be achieved with smaller truncation ratios. However, some of the lenses of Fig. 8 may have to be corrected to avoid excessive aberration due to coma.

In another experiment the system of Fig. 7 was set up for evaluation. The same laser was used as in the previous experiment with the 48x telescope L_1-L_2 enlarging the beam diameter to 3.5 cm. With the physical aperture of the sound cell being 2.5 cm, this resulted in a truncation ratio of 0.71. Cylinder lenses L_3 and L_4 were simple uncorrected lenses of 25 cm focal length. The final telescope L_5-L_6 had a power of 60x consisting of lenses L_5 of 360 mm focal length and L_6 of 6 mm focal length. The frequency swing was 20 MHz. The effective number of resolvable points was determined in a similar way as in the previous experiment. It was found to be 107 at 20 MHz, 112 at 30 MHz, and 107 at 40 MHz. The expected number was 122 based on the truncation ratio of 0.71. The small discrepancy is ascribed to aberration in the final telescope. It should be noted that aberration in the non-scan direction was quite noticeable as is to be expected on the basis of eq. (8).

Efficiency

The theoretical fraction of input light which may be deflected by a Bragg diffraction cell is given by(8)(11)

$$P_d = P_i \sin^2 \left[\frac{\pi^2 P_s}{2\lambda^2} \left(\frac{\ell}{h} \right) M \right]^{\frac{1}{2}} \tag{9}$$

where P_d is the deflected light power, P_i is the input light power, P_s is the sound power, λ is the light wavelength in vacuum, ℓ is the transducer length (along the path of the light beam), and h is the transducer height. The parameter(9)(10)

$$M = \frac{n^6 p^2}{\rho V^3} \tag{10}$$

is an acousto-optic figure of merit dependent on the deflector material's index of refraction n, sound velocity V_s, density ρ and strain-optic coefficient p. Note that, for small values of sound power, P_s, when the sine function above is approximately equal to its argument, the deflected light power is directly proportional to the sound power. It can also be seen that 100% of the input light may theoretically be deflected when the argument of the sine is $\pi/2$ i.e., when

$$P_s = \frac{\lambda^2}{2} \frac{h}{\ell M} \tag{11}$$

Thus, the acoustic power requirement decreases as the transducer height aspect ratio, h/ℓ, gets smaller. This, in fact, is the reason why a sheet sound beam is used. For the glass deflector used in the experiment, $h = 2$ mm, $\ell = 50$ mm, $M = .635\ 10^{-14}\ m^2/\text{watt}$ and $\lambda = .633\ (10^{-6})$m. Calculationsbased on (11) indicate that the maximum light deflection efficiency should occur when $P_s = 1.26$ watts. Experimentally the peak efficiency measured exceeded 90% at an acoustic power level of 1.68 watts when the input light was incident at the proper Bragg angle. Taking into account probable transducer insertion loss, this is in satisfactory agreement with the calculations.

The proper angle of incidence (Bragg angle) is a function of acoustic frequency[11] given by

$$\theta_B = \frac{\lambda f_s}{2V_s} \tag{12}$$

where f_s is the sound frequency. Therefore, in scanning applications where a varying frequency is applied to the cell, the sound beam would ideally have to rotate with frequency so as to insure optimum operation. Although this feature is incorporated in certain high resolution scanners by means of a phased array transducer[6], it is not necessary in the medium resolution scanner used in the experiment. This Bragg angle tolerance, written in terms of the acoustic frequency change δf which gives a drop of 4 db in deflection efficiency, is[12]

$$|\delta f| < \frac{V_s^2}{\lambda f_s \ell} \tag{13}$$

Putting in the values of $f_s = 40$ MHz, and $\ell = 55$ mm for the deflector of Fig. 6 gives a $2\delta f$ (the allowable bandwidth) of 21.6 MHz. Thus, in this deflector, the normal tolerance on Bragg angle incidence is broad enough to cover the relatively small 20 MHz frequency sweep. The percentage of diffracted light as a function of frequency is shown in Fig. 9 [13] for two drive

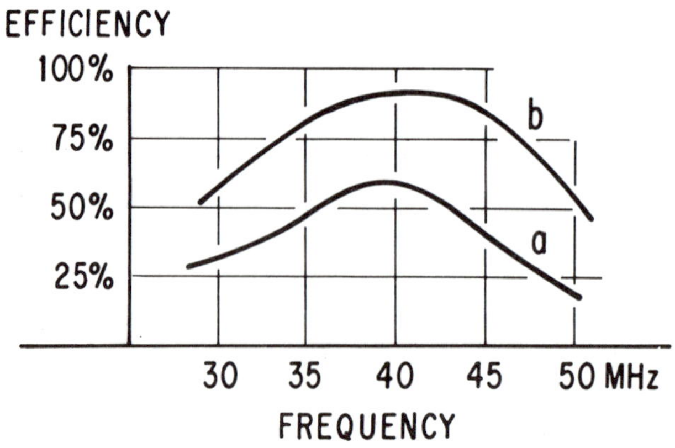

Fig. 9. Efficiency Curves For Deflector

levels; a) 0.5 watts and b) 2.0 watts. The curves are traced from photographs of an oscilloscope screen and show the results of an experiment with an optical configuration as shown in Fig. 8. The electrical power was derived from a sweep generator (Jerrold Model 707) followed by a wide band power amplifier (E.N.I. Model 300L). The output terminal of the amplifier was connected to the deflector via a BNC cable whose electrical length was one-quarter wavelength at the center frequency of 40 MHz. It should be stressed that use of a well-matched properly-designed amplifier is necessary to obtain the uniform response shown in Fig. 9. As seen from this figure, with 2 watts applied, the deflector efficiency is over 90% at the center frequency, dropping to ~50% at the edges.

Access Time

As a random access deflector the device tested is characterized by an access time of approximately 6.5 μsec when used with the full aperture of 25 mm. The access time may be reduced by using a smaller aperture with corresponding loss in resolution. Basically the access time represents the transit time of the sound through the light beam. When using a linear scan access time problems of this kind do not exist. Rather, the frequency "chirp" present in the aperture acts as an extra cylinder lens[7] whose presence may be easily compensated for by readjusting the optical system.

REFERENCES

(1) A. Korpel, R. Adler, P. Desmares and W. Watson, "A Television Display Using Acoustic Deflection and Modulation of Light", Joint Issue of IEEE Proc., Vol. 54 and Applied Optics, Vol. 5, October 1966.

(2) F. A. Jenkins and H. E. White, "Fundamentals of Optics", McGraw-Hill, 3rd edition, 1957, p. 300.

(3) J. Randolph and J. Morrison, "Modulation Transfer Characteristics of an Acousto Optic Deflector", to be published in Applied Optics, June 1971.

(4) J. Randolph and J. Morrison, "Rayleigh-Equivalent Resolution of Acousto Optic Deflection Cells", to be published in Applied Optics, June 1971.

(5) J. W. Goodman, "Introduction to Fourier Optics", McGraw-Hill, Quantum Electronic Series, 1st edition, 1968, p. 114.

(6) #3,424,906 and #3,493,759. U.S. Patents.

(7) G. S. Gerig and H. Montague, "A Simple Optical Filter for Chirp Radar", Proc. IEEE (Correspondence), vol. 52, p. 1753, December 1964.

(8) C. F. Quate, C. D. W. Wilkinson and D. K. Winslow, "Interaction of Light and Microwave Sound", Proc. IEEE, vol. 53, pp. 1604-1673, October 1965.

(9) T. M. Smith and A. Korpel, "Measurement of Light Sound Interactions in Solids", IEEE Jour. of Quantum Electronics, vol. QE-1, Number 6, pp. 283-284, September 1965.

(10) R. W. Dixon and M. G. Cohen, "A New Technique for Measuring Magnitudes of Photoelastic Tensors", App. Phys. Letters 8, Number 8, p. 205, 15 April 1966.

(11) R. Adler, "Interaction Between Light and Sound", IEEE Spectrum, pp. 42-53, May 1969.

(12) W. H. Watson and G. W. Hrbek, "A High Intensity Modulator for Laser Light", Proceedings of the Technical Program, 1970 Electro-optical Systems Design Conference, pp. 630-639, Sept. 22-24, 1970.

(13) This and most of the other figures have been taken from Zenith Application Note "Medium Resolution Scanning with an M-40R Acousto-Optic Modulator".

ACOUSTO-OPTIC RECONSTRUCTION OF MICROWAVE AND ACOUSTIC HOLOGRAMS

Marvin King
Riverside Research Institute
80 West End Avenue
New York, New York 10023

ABSTRACT

A novel technique is proposed for converting an acoustic or microwave hologram into a visible three-dimensional image in real time. The image is formed by coupling the electrical output signals of a planar array of hologram field detectors to a multiple-channel Debye-Sears light modulator, which converts the object-scattered field distribution into an analogous optical-wavelength field distribution instantaneously. The optical radiation propagates in such a way as to reconstruct a visible image of the object.

INTRODUCTION

In order to produce a visible image instantaneously from a microwave or an acoustic hologram, it is necessary to convert the object-scattered wave field to an analogous optical wave distribution. A method of accomplishing this conversion in real time is the central topic of this paper; it is based upon feeding electrical signals derived from a planar array of field detectors into a multiple-channel Debye-Sears light modulator. Two principal limitations of this scheme will be discussed. First, the number of independent resolution cells that can be provided in the reconstructed image is limited by the number of acoustic channels that can be placed into the multiple-channel light modulator. Second, in terms of axial and lateral magnifications, the reconstructed image is likely to be severly distorted because of the disparate aperture sizes and wavelengths used for forming the original hologram and reconstructing the scene. Other limitations that will not be discussed here include the complicated electronic system required to couple the field detector array signals into the light modulator, acoustic wave attenuation in the light modulator, and opto-mechanical problems.

PRINCIPLES OF OPERATION

Assume that the object is illuminated with a source of coherent microwave or acoustic radiation. A portion of the radiation scattered by the object is detected by a square N x N element planar array of field detectors, shown in Fig. 1. The set of electric signal pulses induced in the detector elements by the object-scattered radiation field constitutes the hologram. This ensemble of signals is used to form an analogous optical transparency instantaneously in a two-step process.

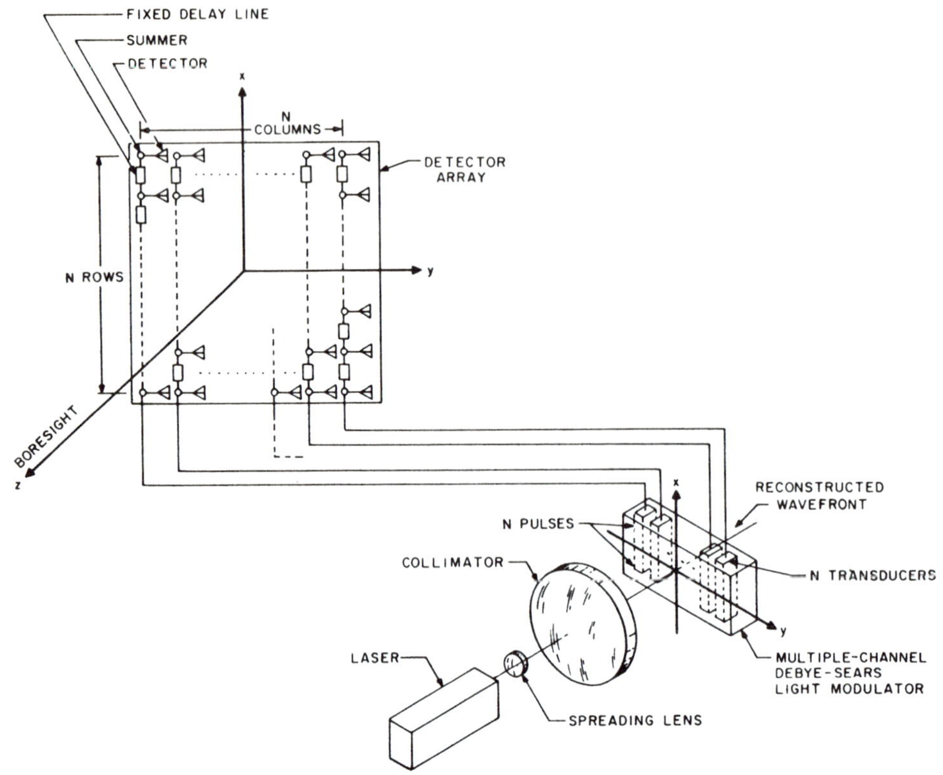

FIG. 1 REAL - TIME HOLOGRAM RECONSTRUCTION

In the first step, the N carrier signals in each column of the array are gated and then multiplexed in time by means of the signal summers and fixed delay lines indicated in the figure. The fixed delay time of the delay lines, which is equal to the pulse repetition period of the time-multiplexed signal, is determined by the design of the Debye-Sears light modulator, to be discussed below. N pulse trains, each containing N pulses, are thus obtained from the detector array. In the second step, this set of signals is connected, after suitable amplification and frequency shifting (not shown), to the N-channel Debye-Sears light modulator.[1] When the signals are completely contained within the light modulator aperture, the ultrasonic disturbance in the delay medium is completely analogous to the object-scattered field detected by the array. This analogy is complete in the sense that a square array of hologram field disturbances has been reproduced by a square array of ultrasonic disturbances where the phase and amplitude of each elementary ultrasonic disturbance is equal to the phase and amplitude of the corresponding hologram field element.

If collimated laser light passes through the Debye-Sears light modulator at this time, the phase of the emerging wavefront is modulated; if the phase modulation is small, it can be shown that the modulated component of the emerging wavefront is the optical analog of the detected amplitude and phase distribution in the planar array. Therefore, this component reconstructs the original scene. An observer may view the reconstructed image by looking into the appropriate first-order diffracted light.

Typically, the image distance from the multiple-channel Debye-Sears light modulator is very large; a practical system is likely to employ a lens, bringing the distant image near to the back focal plane of the lens. In this case, real images are formed.

Note that the image appears only for the time that the N x N signal array remains completely within the Debye-Sears light modulator aperture. Thus, unless the height (or propagation distance) of the modulator is long, the fully reconstructed image appears only for a few microseconds. In order to make a single image visible to a human observer, it would be necessary to repeat the signals at a frame rate of, say, 24 or 30 frames per second to create the illusion of a static image. On the other hand, if the detector array is viewing a moving scene, then this frame rate would be used to illuminate the target and produce an apparent continuum of images to a human observer.

The image, of course, may also be detected by electronic means such as orthicons or vidicons. It should be noted, however, that the optical systems associated with these detectors have limited depth of field and may place this detection approach at a disadvantage for scenes with large depth coverage.

If the target is stationary or slowly moving, then a time-multiplexed signal representing the field in each column of the array may be obtained simply by gating the field detector elements sequentially and summing, obviating the need for delay lines. However, when the target moves rapidly it is necessary to gate all detector elements nearly simultaneously and to combine them with the delay line system described above. Regardless of whether sequential or simultaneous gating is used, the pulse repetition period in the time-multiplexed column signal and the duration of each pulse in the time-multiplexed signal are determined by the requirement of a square format in the array of elementary pulses in the light-modulator aperture; the effective "row" and "column" spacings in the modulator must be equal in order to provide this square format. Therefore, the pulse repetition period must be such that the ultrasonic propagation distance covered in that period of time is equal to the center-to-center spacing of adjacent ultrasonic channels in the light modulator. Furthermore, the pulse duration must be no greater than the pulse repetition period to avoid overlap in the time-multiplexed column signal; it can be shown that pulse overlap in the column signals would create severe vignetting in the reconstructed image.

EXPERIMENTAL DEMONSTRATION

The principles of operation of the proposed reconstruction scheme have been demonstrated[2] with a 24-channel Debye-Sears light modulator used for processing simulated signals received from point targets in the far field of a 24 x 24 element microwave planar array antenna system. It was found that the resolution and sidelobe structure expected from the microwave antenna aperture were preserved in the multiple-channel Debye-Sears light modulator approach.

PERFORMANCE LIMITATIONS

It may be shown that the reconstructed image obtained from this system will be distorted for practical acoustic and microwave hologram signal wavelengths. It is further found that the number of resolution cells in each plane of the reconstructed image is, at most, equal to the number of individual field detector elements that can be processed in the multiple-channel Debye-Sears light modulator.

In order to understand the nature of these limitations, consider the Debye-Sears light modulator shown in Fig. 2.

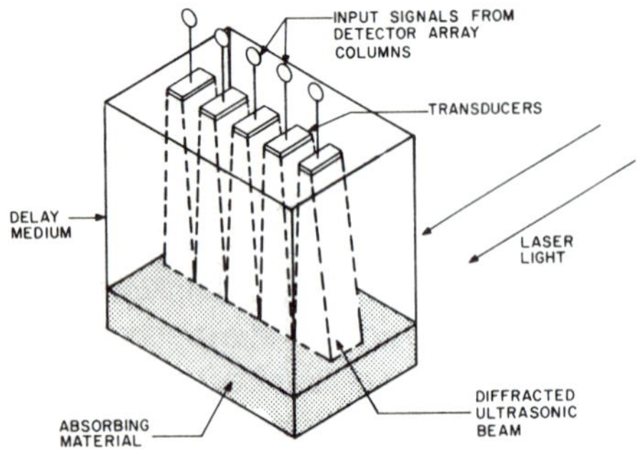

FIG.2 FIVE-CHANNEL DEBYE-SEARS LIGHT MODULATOR

Note that the modulator consists of a square-apertured slab of transparent ultrasonic delay medium terminated with an ultrasonic absorbing material. In the example of this figure, five regularly spaced piezoelectric transducers are bonded to the top surface of the delay medium, forming five channels. (N transducers would be used for an N-column detector array.)

The diffraction spread of the ultrasonic beams limits the maximum number of channels that can be placed in a slab of delay medium of a given size; if the transducers are spaced too closely then the ultrasonic beams overlap and severe vignetting of the reconstruction results. The channel separation may be diminished, however, by increasing the ultrasonic frequency; this effect may be exploited until the frequency is limited either by ultrasonic absorption or by the practicality of using high-frequency equipment. The size of the delay medium is ordinarily limited by the availability of high-quality acousto-optic materials to square apertures of 15 x 15 cm or less.

When ultrasonic diffraction limits the number of channels, it may be shown[2] that the maximum number of channels that can be placed in a square aperture is given by

$$N = \tfrac{1}{2}(f_u T)^{\frac{1}{2}},$$

where f_u is the ultrasonic carrier frequency and T is the ultrasonic transit time of the aperture. It may similarly be shown that the required transducer bandwidth is given roughly by

$$B = N/T = \tfrac{1}{2}(f_u/T)^{\frac{1}{2}}.$$

For a square aperture, the number of array elements that can be processed is simply given by

$$N^2 = f_u T/4.$$

It is seen that the number of array elements is proportional to ultrasonic carrier frequency and modulator transit time. This number, which is also equal to the number of picture elements in each reconstructed image plane, is expected to be limited to a few thousand for practical situations (see below). However, this limited number of resolution elements may yield very fine image resolution over a reduced field of view by appropriate design of the hologram detector array.

The results of calculations on N^2 are shown in the right-hand side of Fig. 3, which is a plot of the number of elements in a square detector array that can be processed as a function of the aperture size for a variety of delay media. It is assumed that the ultrasonic frequency in all of these systems is equal to 500 MHz. The graph shows that a sample of $LiTaO_3$ of size 12 x 12 cm can process a square detector array containing about 2,500 elements, but a slab of $LiNbO_3$ of the same size can only process some 2,200 elements because of its higher ultrasonic propagation speed.

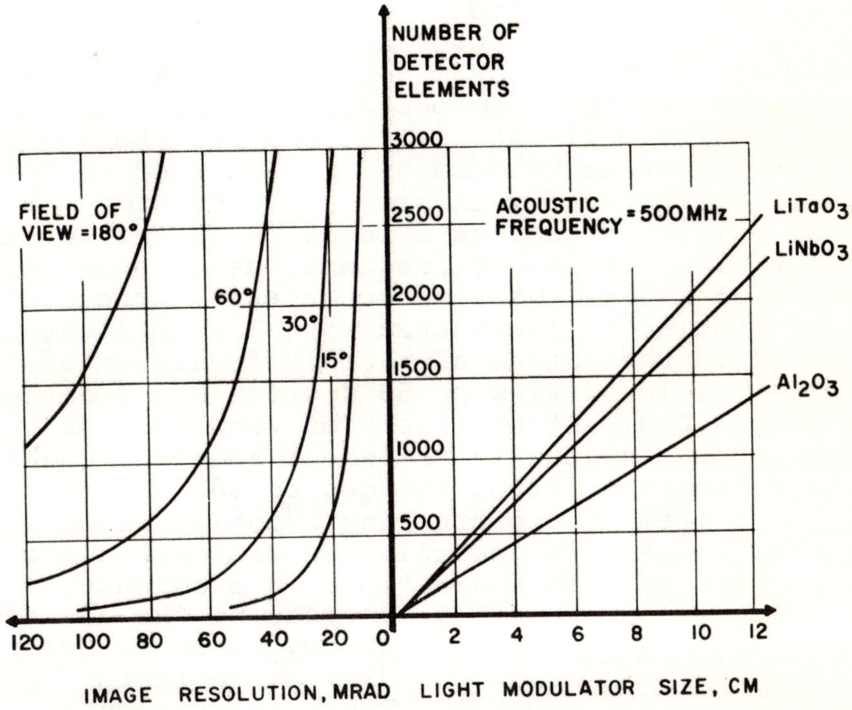

FIG. 3 NUMBER OF DETECTOR ELEMENTS, MODULATOR SIZE, AND IMAGE RESOLUTION

For a fixed number of detector elements, it is possible to obtain very fine resolution over a small field of view by increasing the interelement spacing. The equations relating angular field of view with interelement spacing in a detector array,[3] used in microwave array antenna design for radar systems and astronomy, apply here. The left-hand side of Fig. 3 shows the result of computing the trade-off between field of view and image resolution for fixed numbers of elements in a square detector array. It is seen, for instance, that a 2,500-element square array may be spaced such that a square field of view of 180 degrees on a side is obtained with an image resolution 80 mrad (this is obtained by using an interelement spacing in the array of one-half the hologram signal wavelength). If the field of view is reduced to 60 degrees on a side (by increasing the interelement spacing to one wavelength), the image resolution is refined to 40 mrad. This 2,500-element detector array may achieve a 10-mrad angular resolution in the image by restricting the field of view to 15 degrees on a side. One concludes that although the scheme under consideration can only process a limited number of image elements, the hologram detector array may be designed so as to yield high-resolution images over a small field of view.

The reconstructed scene will be distorted[4] unless the ratio of the hologram wavelength to the detector array aperture is equal to the ratio of the reconstructing optical wavelength to the Debye-Sears light modulator aperture. Define the image distortion factor as the ratio of the axial to lateral magnifications of the reconstructed image; if the reconstructed image is undistorted, then the distortion factor is unity. Since the spacing between ultrasonic channels in the Debye-Sears light modulator is equal to many optical wavelengths, it may be expected that the image distortion factor approaches unity when the interelement spacing of the hologram detector array is also equal to many signal wavelengths; this occurs when the hologram field of view is small. The well-known equations for the magnification of reconstructed images in holography[4] may be used to compute the image distortion factor for this scheme. Figure 4 is a plot of the image distortion factor as a function of the detector array field of view with ultrasonic frequency as a parameter. For this graph, a 35 x 35-element detector array processed by a $LiNbO_3$ light modulator illuminated with 500-nm light has been assumed. Observe that at an ultrasonic frequency of 500 MHz the detector array field of view of 50 degrees has a distortion factor of over 3,000. If the field of view is reduced to 0.5 degrees (by increasing the interelement spacing roughly 100 times), the distortion factor reduces to 30. Further reduction of the field of view to one arc minute can be shown to produce an undistorted image. In general, the distortion factor is inversely proportional to the ultrasonic frequency, so that increasing the ultrasonic frequency to 1,000 MHz reduces the distortion factor by one-half. Thus, it is seen that the image reconstructed by this instantaneous conversion system is expected to be substantially distorted for ordinary optical fields of view. This distortion may be improved by utilizing a hologram detection array with large spacings between the elements, yielding a small detector array field of view.

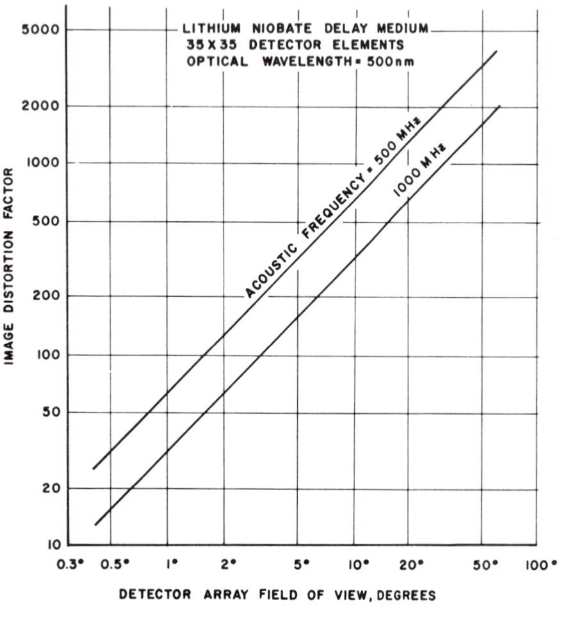

FIG. 4 IMAGE DISTORTION

CONCLUSIONS

A novel system utilizing a multiple-channel Debye-Sears light modulator for reconstructing microwave and acoustic holograms visibly and instantaneously has been described. It may be expected that only a few thousand image resolution elements per depth plane of the field of view may be reconstructed with currently available materials and technology. Except for very small fields of view, the reconstructed images are severely distorted. Finally, the principles of operation of this system have been demonstrated in the processing of signals from microwave planar array antennas utilizing Debye-Sears light modulators.

REFERENCES

1. L.B. Lambert, "Wide-Band, Instantaneous Spectrum Analyzers Employing Delay-Line Light Modulators," IRE Int'l. Conv. Rec. 10, part 6, p.69(1962).
2. L.B. Lambert, M. Arm, and A. Aimette, "Electro-Optical Processors for Phased Array Antennas," in Optical and Electro-Optical Information Processing, J.T. Tippett, et al., Eds. (MIT Press, Cambridge, 1965), p.715.
3. M.I. Skolnik, Introduction to Radar Systems (McGraw-Hill, New York, 1962), p.294.
4. J.A. Armstrong, "Fresnel Holograms: Their Imaging Properties and Aberrations," IBM Journal, May 1965, p.171.

BIBLIOGRAPHY

Entries in this bibliography are arranged alphabetically by the author or first named for co-authored publications.

ACOUSTIC SURFACE WAVES

Adam, J.D., DELAY OF MAGNETOSTATIC SURFACE WAVES IN Y.I.G., Electron Lett v6 n22 p718 (29 Oct 70)

Adkins, L.R., Hughes, A.J., ELASTIC SURFACE WAVES GUIDED BY THIN FILMS: GOLD ON FUSED QUARTZ, IEEE Trans Microwave T&T vMTT-17 n11 p904 (Nov 69)

Adkins, L.R., Hughes, A.J., ACOUSTIC SURFACE-WAVE DIRECTIONAL COUPLER, (Abstr.) 1969 IEEE Ultrasonic Symp, IEEE Trans Sonics & US vSU-17 n1 p61 (Jan 70)

Adkins, L.R., Bristol, T.W., Hagon, P.J., Hughes, A.J., SURFACE ACOUSTIC WAVES - DEVICE APPLICATIONS AND SIGNAL ROUTING TECHNIQUES FOR V.H.F. AND U.H.F., Microwave J v13 p87 (Mch 70)

Adkins, L.R., Hughes, A.J., LONG DELAY LINES EMPLOYING SURFACE ACOUSTIC WAVE GUIDANCE, J Appl Phys v42 n5 p1819 (Apr 71)

Adler, E.L., Sun, I-H. OBSERVATION OF LEAKY RAYLEIGH WAVES ON A LAYERED HALF-SPACE, IEEE Trans Sonics & US vSU-18 n3 p181 (Jly 71)

Adler, R., Korpel, A., Desmares, P., AN INSTRUMENT FOR MAKING SURFACE WAVES VISIBLE, IEEE Trans Sonics & US vSU-15 n3 p157 (Jly 68)

Adler, R., SIMPLE THEORY OF ACOUSTIC AMPLIFICATION, (Abstr.) IEEE Trans Sonics & US vSU-17 n1 p53 (Jan 70)

Adler, R., DeVries, A.J., Kessler, L.W., THE LIQUID WEDGE - AN EFFICIENT SURFACE WAVE TRANSDUCER, (Abstr) IEEE Trans Sonics & US vSU-17 n1 p56 (Jan 70)

Adler, R., SIMPLE THEORY OF ACOUSTIC AMPLIFICATION, IEEE Trans Sonics & US vSU-18 n3 p115 (Jly 71)

Adrion, R.F., THE SO-CALLED DOPPLER SHIFT IN BRAGG CELLS, IEEE Proc v58 n9 p1391 (Sep 70)

Aggarwal, H.R., Ablow, C.M., ON THE RAYLEIGH EQUATION FOR ELASTIC SURFACE WAVES, Ac Soc Am J v47 n5(Pt.2) p1461 (May 70)

Akao, F., ATTENUATION OF ELASTIC SURFACE WAVES IN THIN FILM SUPERCONDUCTING Pb AND In AT 316 MHZ, Phys Lett v30A p409 (1 Dec 69)

Albani, M., Bernardi, P., A FERRITE-METAL SURFACE WAVE IN A RECTANGULAR GUIDE LOADED WITH A FERRITE SLAB PLACED AGAINST A BROAD WALL, IEEE Proc v58 n7 p1175 (Jly 70)

Alldred, G.P., Allen, R.E., De Wette, F.W., CHARACTER OF SURFACE AND PSEUDO-SURFACE WAVES IN A DISCRETE LATTICE, Ac Soc Am J v49 n5(Pt.2) p1453 (May 71)

Altman, L., TAPPING MICROWAVE ACOUSTICS FOR BETTER SIGNAL PROCESSING, Electronics v42 n23 p94 (10 Nov 69)

Anderholm, N.C., LASER GENERATED STRESS WAVES, Appl Phys Lett v16 n3 p113 (1 Feb 70)

Anon, ACOUSTIC SURFACE WAVES DEFLECT LASER BEAM, Electro-optical Syst Des v2 n11 p9 (Nov 70)

Anon, SURFACE ACOUSTIC WAVES USED TO DEFLECT FILM-GUIDED LASER LIGHT, Opt Spectra v4 n10 p11 (Nov 70)

Anon, ACOUSTO VS ELECTRO-OPTICAL MODULATORS, Electro-optical Syst Des v3 n1 p22 (Jan 71)

Armstrong, D.B., SURFACE WAVE TRANSDUCERS FOR DELAY APPLICATIONS IN THE LOW MICROWAVE-FREQUENCY RANGE (Abstr.), IEEE Trans Sonics & US vSU-16 n1 p20 (Jan 69)

Arora, R.K., Vijayaraghanan, S., SCATTERING OF A SHIELDED SURFACE WAVE BY A WALL-IMPEDANCE DISCONTINUITY, IEEE Trans Microwave T&T vMTT-18 n10 p734 (Oct 70)

Arzt, R.M., Dransfeld, K., EXCITATION OF RAYLEIGH WAVES AT HIGH FREQUENCIES AND AT LOW TEMPERATURES, Appl Phys Lett v7 n6 p156 (15 Sep 65)

Arzt, R.M., Salzmann, E., Dransfeld, K., ELASTIC SURFACE WAVES IN QUARTZ AT 316 MHZ, Appl Phys Lett v10 n5 p165 (1 Mch 67)

Ash, E.A., Morgan, D., REALIZATION OF MICROWAVE CIRCUIT FUNCTIONS USING ACOUSTIC SURFACE WAVES, Electron Lett v3 n10 p462 (Oct 67)

Ash, E.A., De La Rue, R.M., Humphryes, R.F., MICROSOUND SURFACE WAVEGUIDES, IEEE Trans Microwave T&T vMtt-17 n11 p882 (Nov 69)

Ash, E.A., TAPERED GRATING REFLECTORS, IBM Tech Discl Bull v13 n5 p1204 (Oct 70)

Atley, E., SURFACE-WAVE DEVICES OFFER CHEAP SIGNAL PROCESSING, Electronic Des p25 (6 Dec 70)

Atzeni, C., Masotti, L., MINIMIZATION OF THE SENSOR NUMBER IN SURFACE WAVE INTERDIGITAL ARRAYS (Abstr.), IEEE Trans Sonics & US vSU-18 n1 p60 (Jan 71)

Auld, B.A., Collins, J.H., Gerard, H.M., Lakin, K.M., Shaw, H.J., Smith, W.R., AURFACE WAVE TRANSDUCERS (Abstr.), IEEE Trans Sonics & US vSU-16 n1 p20 (Jan 69)

Auld, B.A., Webb, D.C., Winslow, D.K., ACOUSTIC BEAM SCANNING BY OPTICAL CONTROL OF PIEZOELECTRIC COUPLING, IEEE Proc v57 n4 p713 (Apr 69)

Auld, B.A., SURFACE WAVE THEORY (Bstr.), IEEE Trans Sonics & US vSU-18 n1 p44 (Jan 71)

Auld, B.A., Wilson, D.A., Winslow, D.K., Young, E., Jr., CONTROL OF ACOUSTIC SURFACE WAVES WITH PHOTOCONDICTIVE CdS FILM, Appl Phys Lett v18 n8 p339 (15 Apr 71)

Auth, D.C., Mayer, W.G., SCATTERING OF LIGHT REFLECTED FROM ACOUSTIC SURFACE WAVES IN ISOTROPIC SOLIDS, J Appl Phys v38 n13 p5138 (Dec 67)

Auth, D.C., NEW HIGH POWER SOURCE OF COHERENT MICROWAVE PHONONS, Appl Phys Lett v16 n12 p521 (14 Jne 70)

Aviram, A., Kuhn, L., MULTI-MONOMOLECULAR FILM MATERIALS FOR ACOUSTIC AND OPTICAL DEVICES, IBM Tech Discl Bull v14 n1 p290 (Jne 71)

Bahr, A.J., Court, I.N., DETERMINATION OF THE ELECTROMECHANICAL COUPLING COEFF. OF THIN-FILM CADMIUM SULPHIDE, J Appl Phys v39 n6 p2863 (May 68)

Bahr, A.J., ELECTRICALLY CONTROLLED REFLECTION OF ACOUSTIC SURFACE WAVES (Abstr.) IEEE Trans Sonics & US vSU-17 n1 p69 (Jan 70)

Baibakov, V.I., ACOUSTIC NOISE IN CdSe, Soviet Phys - Solid State, v9 n7 p1654 (Jan 68)

Baibakov, V.I., ACOUSTIC NOISE AND CURRENT SATURATION IN CdSe, Soviet Phys - Semiconductors v2 n1 p128 (Jly 68)

Barrekette, E.S., Lean, E.G., PATTERN RECOGNITION BY SURFACE ACOUSTIC WAVES, IBM Tech Discl Bull v13 n12 p3599 (May 71)

Bateman, T.B., SOME MATERIALS FOR ULTRASONIC TRANSDUCER BONDING AT CRYOGENIC TEMPERATURES, Ac Soc Am J v41 n4 p1011 (67)

Bechmann, R., ELASTIC AND PIEZOELECTRIC CONSTANTS OF ALPHA-QUARTZ, Phys Rev v110 n5 p1060 (1 Jne 58)

Bennett, G.A., Adam, J.D., IDENTIFICATION OF SURFACE-WAVE RESONANCES ON A METAL-BACKED Y.I.G. SLAB, Electron Lett v6 n24 p789 (26 Nov 70)

Berlincourt, D.A., Curran, D.R., Jaffe, H., Physical Acoustics, W.P. Mason, ed., v1A New York: Academic Press, p233 (1964)

Berlincourt, D.A., PIEZOELECTRIC AND FERROELECTRIC ENERGY CONVERSION, IEEE Trans Sonics & US vSU-15 n2 p89 (Apr 68)

Berlincourt, D.A., PIEZOELECTRIC TRANSDUCERS, Electro-Technology, Hardware Report, p33 (Jan 70)

Bernstein, B.T., ELASTIC CONSTANTS OF SYNTHETIC SAPPHIRE AT 27°C, J Appl Phys v34 n1 p169 (Jan 63)

Bers, A., Burke, B.E., RESONANT AMPLIFICATION AND COUPLING OF ACOUSTIC SURFACE WAVES WITH ELECTRONS DRIFTING ACROSS A MAGNETIC FIELD, Appl Phys Lett v16 n8 p300 (15 Apr 70)

Bers, A., INTERACTIONS BETWEEN ACOUSTIC SURFACE WAVES AND ELECTRONS IN SOLIDS (Abstr.), IEEE Trans Sonics & US vSU-18 n1 p53 (Jan 71)

Bertoni, H.L., PIEZOELECTRIC RAYLEIGH WAVE EXCITATION BY BULK WAVE SCATTERING, IEEE Trans Microwave T&T vMTT-17 p873 (Nov 69)

Bharat, R., Zaidi, M.A., ULTRASONIC SURFACE-WAVE DELAY LINES USING COMPOSITE SOLID-FLUID STRUCTURES (Abstr.), IEEE Trans Sonics &US vSU-18 n1 p67 (Jan 71)

Blackledge, V.O., Kaufman, I., CONTROL OF ELECTRON EMISSION BY ELASTIC SURFACE WAVES, Appl Phys Lett v15 p127 (Aug 69)

Blackledge, V.O., Sirkis, M.D., Kaufman, I., INVESTIGATION OF DISPLAY TECHNIQUES USING ELASTIC SURFACE WAVES AND LOCAL STORAGE, IEEE J Solid State Circuits vSC-5 n5 p244 (Oct 70)

Bleustein, J.L., A NEW SURFACE WAVE IN PIEZOELECTRIC MATERIALS, Appl Phys Lett v13 p412 (Dec 68)

Blotekjaer, K., Quate, C.F., THE COUPLED MODES OF ACOUSTIC WAVES AND DRIFTING CARRIERS IN PIEZOELECTRIC CRYSTALS, IEEE Proc v52 n4 p360 (Apr 64)

Bogdanov, S.V., Yakovkin, I.B., ATTENUATION OF ELASTIC SURFACE WAVES IN A SUBSTRATE - InSb FILM SYSTEM, Soviet Phys - Semiconductors v3 n4 p500 (Oct 69)

Bond, W.L., Collins, J.H., Gerard, H.M., Reeder, T.M., Shaw, H.J., ACOUSTIC SURFACE-WAVE COUPLING ACROSS AN AIR GAP, Appl Phys Lett v14 p122 (Feb 69)

Bond, W.L., Reeder, T.M., Shaw, H.J., WRAP-AROUND SURFACE-WAVE DELAY LINES, Electron Lett v7 n3 p79 (11 Feb 71)

Bongianni, W.L., PULSE COMPRESSION USING NONLINEAR INTERACTION IN A SURFACE ACOUSTIC WAVE CONVOLVER, IEEE Proc v59 n4 p713 (Apr 71)

Bradfield, G., SURFACE WAVE VELOCITY MEASUREMENT, Nature v211 n5051 p840 (20 Aug 66)

Bradfield, G., DAMPING OF SURFACE (SOUND) WAVES BY THE AIR, Electron Lett v4 n5 p95 (8 Mch 68)

Broers, A.N., Lean, E.G., Hatzakis, M., 1.75 GHz ACOUSTIC SURFACE WAVE TRANSDUCER FABRICATED BY AN ELECTRON BEAM, Appl Phys Lett v15 n3 p98 (1 Aug 69)

Brundle, L.K., Freedman, N.J., MAGNETOSTATIC SURFACE WAVES ON A Y.I.G. SLAB, Electron Lett v4 n7 p132 (Apr 68) Erratum: v4 n5 p182 (May 68)

Bruun, M., Ludvik, S., Quate, C.F., FIELD EFFECT TRANSISTORS ON EPITAXIAL GaAs TRANSDUCERS FOR ACOUSTIC SURFACE WAVES, Appl Phys Lett v18 n4 p119 (15 Feb 71)

Buchwald, V.T., RAYLEIGH WAVES IN TRANSVERSELY ISOTROPIC MEDIA, Quaterly J of Mechanics and Appl Math v14(Pt.3) p293 (Aug 61)

Buchwald, V.T., Davis, A., SURFACE WAVES IN ANISOTROPIC ELASTIC MEDIA, Nature v191 n4791 p899 (Aug 26, 61)

Budreau, A.J., Carr, P.H., SURFACE-WAVE ATTENUATION AT MICROWAVE FREQUENCIES: PART II - TEMPERATURE DEPENDENCE, IEEE Trans Sonics & US vSU-18 n1 p50 (Jan 71)

Budreau, A.J., Carr, P.H., TEMPERATURE DEPENDENCE OF THE ATTENUATION OF MICROWAVE FREQUENCY ELASTIC SURFACE WAVES IN QUARTZ, Appl Phys Lett v18 n6 p239 (15 Mch 71)

Bunney, R.E., Goodman, R.R., Marshall, S.W., RAYLEIGH AND LAMB WAVES ON CYLINDERS, Ac Soc Am J v46 n5(Pt.2) p1223 (69)

Burke, R.E., Bers, A., Smith, H.I., Cohen, R.A., Mountain, R.W., ACOUSTIC SURFACE WAVE AMPLIFICATION USING AN ACCUMULATION LAYER ON SILICON, IEEE Proc v58 n10 p1775 (Oct 70)

Burke, B.E., SURFACE-WAVE AMPLIFICATION USING AN ACCUMULATION LAYER ON SILICON (Abstr,), IEEE Trans Sonics & US vSU-18 n1 p54 (Jan 71)

Burnsweig, J., Gregory, E.H., Wagner, R.J., SURFACE-WAVE DEVICE APPLICATIONS AND COMPONENT DEVELOPMENTS, IEEE J Solid State Circuits vSC-5 n6 p310 (Dec 70)

Cachier, G., OPTICAL EXCITATION OF HIGH-AMPLITUDE SURFACE WAVES, Appl Phys Lett v17 n10 p419 (15 Nov 70)

Cambon, G., Quate, C.F., DISPERSIVE RAYLEIGH WAVES ON QUARTZ, Electron Lett v5 n17 p402 (21 Aug 69)

Cambon, G., Rouzeyre, M., ATTENUATION OF DISPERSIVE RAYLEIGH WAVES ON QUARTZ, Electron Lett v6 n17 p539 (20 Aug 70)

Cambon, G., Rouzeyre, M., Simon, G., OPTICAL PROBING OF SURFACE RAYLEIGH WAVES, Appl Phys Lett v18 n7 p295 (1 Apr 71)

Campbell, J.J., Jones, W.R., A METHOD FOR ESTIMATING OPTIMAL CRYSTAL CUTS AND PROPAGATION DIRECTIONS FOR EXCITATION OF PIEZOELECTRIC SURFACE WAVES, IEEE Trans Sonics & US vSU-15 n4 p209 (Oct 68)

Campbell, J.J., Jones, W.R., NUMERICAL STUDY OF ACOUSTIC SURFACE WAVE PROPAGATION IN PIEZOELECTRIC SUBSTRATES, (Abstr.) IEEE Trans Sonics & US vSU-16 n1 p23 (Jan 69)

Campbell, J.J., Jones, W.R., PROPAGATION OF SURFACE ACOUSTIC WAVES AT THE BOUNDARY BETWEEN A PIEZOELECTRIC CRYSTAL AND A FLUID MEDIUM, IEEE Trans Sonics and US vSU-17 n2 p71 (Apr 70)

Campbell, J.J., Jones, W.R., PROPAGATION OF PIEZOELECTRIC SURFACE WAVES ON CUBIC AND HEXAGONAL CRYSTALS, J Appl Phys v41 n7 p2796 (Jne 70)

Carr, P.H., MEASUREMENT OF THE PIEZOELECTRIC CONSTANT OF QUARTZ AT GIGACYCLE FREQUENCIES, Ac Soc Am J v41 n1 p75 (Jan 67)

Carr, P.H., THE GENERATION AND PROPAGATION OF ACOUSTIC SURFACE WAVES AT MICROWAVE FREQUENCIES, IEEE Trans Microwave T&T vMTT-17 n11 p845 (Nov 69)

Carr, P.H., MULTIPLY TAPPED SURFACE-WAVE DELAY LINES AT MICROWAVE FREQUENCIES, (Abstr.), IEEE Trans Sonics & US vSU-17 n1 p61 (Jan 70)

Carr, P.H., INTERACTION BETWEEN NONCOLLINEAR RAYLEIGH WAVES (Abstr.), IEEE Trans Sonics & US vSU-18 n1 p51 (Jan 71)

Carr, P.H., Franklin, S.B., Slobodnik, A.J. Jr., Szabo, T.L., DeVito, P.A., Bush, H.F., Entzminger, J.N., USES AND LIMITATIONS OF MULTIPLY TAPPED SURFACE-WAVE DELAY LINES (Abstr.), IEEE Trans Sonics & US vSU-18 n1 p62 (Jan 71)

Čermak, I., Silvester, P., Farnell, G.W., CAPACITY AND FIELD DISTRIBUTION FOR INTERDIGITAL SURFACE-WAVE TRANSDUCERS, (Abstr.), IEEE Trans Sonics & US vSU-17 n1 p55 (Jan 70)

Chang, I., SATURATION OF ACOUSTIC AMPLIFICATION IN PIEZOELECTRIC SEMICONDUCTORS, IEEE Trans Sonics & US vSU-16 n2 p60 (Apr 69)

Chao, G., PARAMETRIC AMPLIFICATION OF SURFACE ACOUSTIC WAVES, Appl Phys Lett v16 n10 p399 (15 May 70)

Chicotka, R.J., Dakks, M.L., Lean, E.G., Scott, B.A., INTEGRATING ACOUSTIC SURFACE WAVE AND SILICON TRANSISTOR TECHNOLOGY, IBM Tech Discl Bull v13 n11 p3223 (Apr 71)

Cho, F.Y., Hunsinger, B.J., Lawson, R.L., SURFACE WAVE COUPLING THROUGH A FLUID INTERFACE, Appl Phys Lett v16 n11 p441 (1 Jne 70)

Cho, F.Y., Lawson, R.L., Hunsinger, B.J., A NOTE ON THE "DETERMINATION OF THE DIFFRACTION PATTERN OF INTERDIGITAL SURFACE-WAVE DEVICES," IEEE Trans Sonics & US vSU-17 n3 p199 (Jly 70)

Cho, F.Y., Peterson, K.M., Hunsinger, B.J., METHOD TO SEPARATE THE DESIRED RAYLEIGH WAVES FROM SPURIOUS ACOUSTIC MODES, Appl Phys Lett v17 n7 p276 (10 Oct 70)

Cho, F.Y., Hunsinger, B.J., Lawson, R.L., SURFACE-WAVE RECIRCULATION ON A PIEZOELECTRIC SUBSTRATE, (Abstr.) IEEE Trans Sonics & US vSU-18 n1 p48 (Jan 71)

Cho, F.Y., Hunsinger, B.J., Lawson, R.L., SURFACE WAVES CIRCULATING ON PIEZOELECTRIC SUBSTRATES, Appl Phys Lett v18 n7 p298 (1 Apr 71)

Chow, K.K., Leonard, W.B., EFFICIENT OCTAVE-BANDWIDTH MICROWAVE LIGHT MODULATORS, IEEE J Quantum Electronics vQE-6 n12 p789 (Dec 70)

Chung, D.H., Simmons, G., ELASTIC WAVE VELOCITIES IN APOLLO 11 AND 12 SAMPLES AND SURFACE-WAVE DISPERSION ON THE MOON (Abstr.), IEEE Trans Sonics & US vSU-18 n1 p50 (Jan 71)

Claiborne, L.T., Jones, W.S., SURFACE-WAVE MATCHED-FILTER EVALUATION (Abstr.), IEEE Trans Sonics & US vSU-17 n1 p61 (Jan 70)

Claiborne, L.T., Staples, E.J., Harris, J.L., PROGRAMMABLE SURFACE WAVE MATERIAL FILTERS USING PIEZORESISTIVE DETECTOR ARRAYS (Abstr.), IEEE Trans Sonics & US vSU-18 n1 p61 (Jan 71)

Coldren, L.A., Kino, G.S., MONOLITHIC ACOUSTIC SURFACE-WAVE AMPLIFIER, Appl Phys Lett v18 n7 p317 (15 Apr 71)

Collins, J.H., Gerard, H.M., Lakin, K.M., Shaw, H.J., 100MHz QUARTZ DELAY LINE UTILIZING RAYLEIGH WAVES, IEEE Proc v56 n9 p1635 (Sep 68)

Collins, J.H., Gerard, H.M., Shaw, H.J., HIGH-PERFORMANCE LITHIUM NIOBATE ACOUSTIC SURFACE WAVE TRANSDUCERS AND DELAY LINES, Appl Phys Lett v13 n9 p312 (1 Nov 68)

Collins, J.H., Lakin, K.M., Quate, C.F., Shaw, H.J., AMPLIFICATION OF ACOUSTIC SURFACE WAVES WITH ADJACENT SEMICONDUCTOR AND PIEZOELECTRIC CRYSTALS, Appl Phys Letters v13 n9 p314 (1 Nov 68)

Collins, J.H., THE STATUS AND FUTURE OF MICROSOUND TECHNOLOGY, IEEE Int'l Conv Digest, New York, p88 (Mch 69)

Collins, J.H., Gerard, H.M., Reeder, T.M., Shaw, H.J., UNIDIRECTIONAL SURFACE WAVE TRANSDUCER, IEEE Proc v57 n5 p833 (May 69)

Collins, J.H., Hagon, P.J., APPLYING SURFACE WAVE ACOUSTICS, Electronics p97 (10 Nov 69)

Collins, J.H., Hagon, P.J., AMPLIFYING ACOUSTIC SURFACE WAVES, Electronics p102 (8 Dec 69)

Collins, J.H., Reeder, T.M., ACTIVE SURFACE ACOUSTIC WAVE DELAY LINES (Abstr.), IEEE Trans Sonics & US vSU-17 n1 p53 (Jan 70)

Collins, J.H., Hagon, P.J., SURFACE WAVE DELAY LINES PROMISE FILTERS FOR RADAR, FLAT TUBES FOR TELEVISION, AND FASTER COMPUTERS, Electronics p110 (19 Jan 70)

Collins, J.H., SURFACE WAVES: THE NEW ELECTRONICS, New Scientist p245 (30 Jly 70)

Collins, J.H., Hagon, P.J., Pulliam, G.R., EVALUATION OF NEW SINGLE CRYSTAL PIEZOELECTRIC MATERIALS FOR SURFACE ACOUSTIC-WAVE APPLICATIONS, Ultrasonics v8 n4 p219 (Oct 70)

Cook, E.G., Van Valkenburg, H.E., SURFACE WAVES AT ULTRASONIC FREQUENCIES, Bull Am Soc Testing Materials, v198 p81 (May 54)

Coquin, G.A., Tiersten, H.F., ANALYSIS OF THE EXCITATION AND DETECTION OF PIEZO ELECTRIC SURFACE WAVES IN QUARTZ BY MEANS OF SURFACE ELECTRODES, Ac Soc Am J v41 n4(Pt.2) p921 (Apr 67)

Coquin, G.A., Dean, R.E., LOW-LOSS EXCITATION OF SURFACE-WAVE WITH WEDGE-TYPE MODE CONVERTERS, (Abstr.), IEEE Trans Sonics & US vSU-18 n1 p52 (Jan 71)

Costanza, S.T., Hagon, P.J., MacNevin, L.A., ANALOG MATCHED FILTERS USING TAPPED ACOUSTIC SURFACE WAVE DELAY LINE, IEEE Trans Microwave T&T vMTT-17 n11 p1042 (Nov 69)

Court, I.N., MICROWAVE ACOUSTIC DEVICES FOR PULSE COMPRESSION FILTERS, IEEE Trans Microwave T&T vMTT-17 p968 (Nov 69)

Dakss, M.L., Kuhn, L., Pennington, K.S., GENERATION OF SURFACE WAVE NETWORKS BY OPTICAL DAMAGE, IBM Tech Discl Bull v12 n11 p1942 (Apr 70)

Dakss, M.L., Lean, E.G., Schoenwald, J., EXCITATION OF ACOUSTIC SURFACE WAVES WITH A GRATING-COUPLED GUNN OSCILLATOR, Appl Phys Lett v18 n4 p137 (15 Feb 71)

Damon, R.W., Eshbach, J.R., MAGNETOSTATIC MODES OF A FERROMAGNETIC SLAB, J Appl Phys v31 p1048 (May 60)

Damon, R.W., Worley, J.C., ACOUSTIC SURFACE-WAVE PHENOMENA AND APPLICATIONS, 1970 Int'l Solid-State Conf. p128

Daniel, M.R., deKlerk, J., Jones, C.K., Patterson, A., SURFACE WAVE PROPAGATION IN CRYSTALLINE QUARTZ, (Abstr.), IEEE Trans Sonics & US vSU-16 n1 p23 (Jan 69)

Daniel, M.R., deKlerk, J., SOME ELECTRICAL AND ACOUSTIC CHARACTERISTICS OF INTERDIGITAL SURFACE WAVE TRANSDUCERS, (Abstr.), IEEE Trans Sonics & US vSU-17 n1 p55 (Jan 70)

Daniel, M.R., deKlerk, J., TEMPERATURE-DEPENDENT ATTENUATION OF ULTRASONIC SURFACE WAVES IN QUARTZ, Appl Phys Lett v16 n1 p30 (1 Jan 70)

Daniel, M.R., deKlerk, J., VELOCITY MEASUREMENTS OF ELASTIC SURFACE WAVES IN THE LAYERED SYSTEM InS ON Al_2O_3, Appl Phys Lett v16 n9 p331 (1 May 70)

Daniel, M.R., deKlerk, J., AN APPARATUS FOR MEASURING THE VELOCITY AND ENERGY FLOW DIRECTION OF ULTRASONIC SURFACE WAVES, Rev Sci Instr v41 n10 p1463 (Oct 70)

Daniel, M.R., deKlerk, J., APPARATUS FOR MEASURING THE VELOCITY AND ENERGY FLOW DIRECTIONS OF SURFACE WAVES (Abstr.) IEEE TRans Sonics & US vSU-18 n1 p66 (Jan 71)

Das, P., Araghi, M.N., ELASTIC SURFACE WAVES IN THE PRESENCE OF FLUID LAYER, Appl Phys Lett v16 n8 p293 (15 Apr 70)

De Jong, M., THICKNESS VIBRATORS OF PIEZOCERAMIC MATERIALS, IEEE Trans Sonics & US vSU-17 n1 p69 (Jan 70)

deKlerk, J., Daniel, M.R., MICROWAVE ACOUSTIC PROPAGATION IN LAYERED PIEZO-ELECTRIC MEDIA (Abstr.) IEEE Trans Sonics & US vSU-17 n1 p58 (Jan 70)

deKlerk, J., Daniel, M.R., MEASURED ELECTRICAL CHARACTERISTICS OF INTERDIGITAL SURFACE WAVE TRANSDUCERS ON LITHIUM NIOBATE, Appl Phys Lett v16 n5 p219 (1 Mch 70)

deKlerk, J., ULTRASONIC TRANSDUCERS PT. 3. SURFACE WAVE TRANSDUCERS, Ultrasonics v9 n1 p35 (Jan 71)

deKlerk, J., MATERIALS FOR SURFACE WAVE APPLICATIONS (Abstr.) IEEE Trans Sonics & US vSU-18 n1 p44 (Jan 71)

Deresiewicz, H., Mindlin, R.D., WAVES ON THE SURFACE OF A CRYSTAL, J Appl Phys v28 n6 p669 (Jne 57)

DeVries, A.J., Adler, R., Dias, J.F., Wojcik, I.J., REALIZATION OF A 40 MHz COLOR TELEVISION I-F RESPONSE USING SURFACE-WAVE TRANSDUCERS ON LEAD ZIRCONATE TITANATE (Abstr.), IEEE Trans Sonics & US vSU-17 n1 p61 (Jan 70)

DeVries, A.J., Subramanian, S., SPURIOUS BULK MODES IN P.Z.T. SURFACE WAVE FILTERS (Abstr.), IEEE Trans Sonics & US vSU-18 n1 p62 (Jan 71)

DeWames, R.E., Wolfram, T., A NEW TYPE OF SURFACE SPIN WAVE IN FERROMAGNETIC FILMS, Appl Phys Lett v15 n9 p297 (1 Nov 69)

DeWames, R.E., Wolfram, T., MAGNETOSTATIC SURFACE MODES IN AN AXIALLY MAGNETIZED ELLIPTICAL CYLINDER, Appl Phys Lett v16 n8 p305 (15 Apr 70)

Dewar, W.J., Beal, J.C., COAXIAL-SLOT SURFACE WAVE LAUNCHERS, IEEE Trans Microwave T&T vMTT-18 n8 p449 (Aug 70)

Diachok, O.I., Hallermeier, R.J., Mayer, W.G., MEASUREMENT OF ULTRASONIC WAVE VELOCITY AND ABSORPTIVITY ON SINGLE-CRYSTAL COPPER, Appl Phys Lett v17 n7 p289 (1 Oct 70)

Diachok, C.I., Mayer, W.G., CRYSTAL ORIENTATION BY LONGITUDINAL/SURFACE-WAVE MODE CONVERSION (Abstr.), IEEE Trans Sonics & US vSU-18 n1 p56 (Jan 71)

Dieulesaint, E., P. Hartmann, ACOUSTIC SURFACE-WAVE FILTERS (Abstr.), IEEE Trans Sonics & US vSU-18 n1 p61 (Jan 70)

Dransfeld, K., KILOMEGACYCLE ULTRASONICS, Sci Am v208 p60 (Apr 63)

Eberhardt, F.J., Andrews, F.A., LASER HETERODYNE SYSTEM FOR MEASUREMENT AND ANALYSIS OF VIBRATION, Ac Soc Am J v48 n3(Pt.1) p603 (Sep 70)

Engan, H., Ingebrigtsen, K.A., Tonning, A., ELASTIC SURFACE WAVES IN ALPHA-QUARTZ: OBSERVATION OF LEAKY SURFACE WAVES, Appl Phys Lett v10 n11 p311 (1 Jne 67)

Engan, H., Hanebrekke, H., Ingebrigtsen, K.A., Jergan, E., NUMERICAL CALCULATIONS ON SURFACE WAVES IN PIEZOELECTRICS, Appl Phys Lett v15 n8 p239 (15 Oct 69)

Engan, H., EXCITATION OF ELASTIC SURFACE WAVES BY SPATIAL HARMONICS OF INTER-DIGITAL TRANSDUCERS, IEEE Trans Electron Dev vED-16 n12 p1014 (Dec 69)

Epstein, M., Joshi, S.G., Serafin, R.J., Vanden Heuvel, A.P., ELECTRON BEAM SENSING OF SURFACE ELASTIC WAVES, IEEE Proc v58 n2 p253 (Feb 70)

Eshbach, J.R., Damon, R.W., SURFACE MAGNETOSTATIC MODES AND SURFACE SPIN WAVES, Phys Rev v118 n5 p1208 (1 Jne 60)

Ewing, W.M., Jardetzky, W.S., Press, F., "ELASTIC WAVES IN LAYERED MEDIA," McGraw-Hill, New York, 1957

Fang, F.F., Lean, E.G., Yu, H.N., SURFACE ACOUSTIC WAVE AMPLIFICATION ON GaAs, IBM Tech Discl Bull v12 n10 p1702 (Mch 70)

Fang, F.F., Keyes, R.W., LAUNCHING ELASTIC SURFACE WAVE ON NONPIEZOELECTRIC MATERIALS, IBM Tech Discl Bull v13 n10 p3022 (Mch 71)

Farnell, G.W., Lim, T.C., PROPERTIES OF ELASTIC SURFACE WAVES, (Abstr.), IEEE Trans Sonics & US vSU-16 n1 p20 (Jan 69)

Farnell, G.W., PROPERTIES OF ELASTIC SURFACE WAVES, in "Physical Acoustics," v6, W.P. Mason, and R.N. Thurston, eds., New York: Academic Press, 1970

Farnell, G.W., Cermak, I.A., Silvester, P., Wang, S.K., CAPACITANCE AND FIELD DISTRIBUTION FOR INTERDIGITAL SURFACE WAVE TRANSDUCERS, IEEE Trans Sonics & US vSU-17 n3 p188 (Jly 70)

Farnell, G.W., SYMMETRY CONSIDERATIONS FOR ELASTIC LAYER MODES PROPAGATING IN ANISOTROPIC PIEZOELECTRIC CRYSTALS, IEEE Trans Sonics & US vSU-17 n4 p229 (Oct 70)

Fischler, C., Yando, S., AMPLIFICATION OF ACOUSTIC WAVES IN A COUPLED SEMICONDUCTOR PIEZOELECTRIC CERAMIC SYSTEM, Appl Phys Lett v13 n10 p351 (15 Nov 68)

Fischler, C., Yando, S., AMPLIFICATION OF GUIDED ELASTIC WAVES IN PIEZOELECTRIC PLATES THROUGH ELECTRICAL COUPLING TO A SEMICONDUCTOR, Appl Phys Lett v15 n11 p366 (1 Dec 69)

Fischler, C., ACOUSTIC ELECTRIC AMPLIFICATION IN COMPOSITE PIEZOELECTRIC AND SEMICONDUCTING STRUCTURES, IEEE Trans Electron Dev vED-17 n3 p214 (Mch 70)

Foster, N.F., THE DEPOSITION AND PIEZOELECTRIC CHARACTERISTICS OF SPUTTERED LITHIUM NIOBATE FILMS, J Appl Phys v40 n1 p420 (Jan 69)

Freedman, N.J., Brundle, L.K., NONLINEAR BEHAVIOR OF MAGNETOSTATIC SURFACE WAVES, Electron Lett v4 p427 (Oct 68)

Gallawa, R.L., PROPAGATION ON A CYLINDRICAL SURFACE WAVE STRUCTURE HAVING RADIALLY NONUNIFORM DIELECTRIC, IEEE Trans Microwave T&T vMTT-18 n8 p432 (Aug 70)

Gazis, D.C., Herman, R., Wallis, R.F., SURFACE ELASTIC WAVES IN CUBIC CRYSTALS, Phys Rev v119 n2 p533 (15 Jly 60)

Gerard, H.M., ACOUSTIC SCATTERING PARAMETERS OF THE ELECTRICALLY LOADED INTERDIGITAL SURFACE WAVE TRANSDUCER,IEEE Trans Microwave T&T vMTT-17 n11 p1045 (Nov 69)

Gerard, H.M., Smith, W.R., DESIGN OF BROADBAND SURFACE-WAVE TRANSDUCERS USING AN EQUIVALENT CIRCUIT MODEL (Abstr.), IEEE Trans Sonics & US vSU-18 n1 p52 (Jan 71)

Gloersen, P.G., White, R.M., BOLOMETRIC DETECTION OF SURFACE ELASTIC WAVES (Abstr.), IEEE Trans Sonics & US vSU-18 n1 p53 (Jan 71)

Gold, L., RAYLEIGH-WAVE PROPAGATION ON ANISOTROPIC (CUBIC) MEDIA, Phys Rev v104 n6 p1532 (Dec 15, 1956)

Grace, M.I., Kedzie, R.W., Kestigian, M., Smith, A.B., ELASTIC WAVE ATTENUATION IN LITHIUM NIOBATE, Appl Phys Lett v9 n4 p155 (15 Aug 66)

Grasse, C.L., Gandolfo, D.A., 400 MHz ACOUSTIC SURFACE-WAVE PULSE EXPANSION AND COMPRESSION FILTER, IEEE Trans Microwave T&T vMTT-19 n6 p558 (Jan 71)

Gregson, V.G., ADAPTATION OF ELASTIC SURFACE WAVE COMPUTER CODES USED IN SEISMOLOGY TO PROBLEMS IN MICROWAVE ACOUSTICS (Abstr.), IEEE Trans Sonics & US vSU-16 n1 p23 (Jan 69)

Gulyaev, Y.V., Pustovoit, V.I., AMPLIFICATION OF SURFACE WAVES IN SEMICONDUCTORS, Soviet Phys - JETP v20 n6 p1508 (Jne 65)

Gulyaev, Y.V., ON THE NONLINEAR THEORY OF ULTRASOUND AMPLIFICATION IN SEMICONDUCTORS, IEEE Trans Sonics & US vSU-17 n2 p111 (Apr 70)

Gulyaev, Y.V., Kmita, A.M., Kotelyanskii, I.M., Medved, A.V., Tursunov, S.S., EXCITATION AND AMPLIFICATION OF SURFACE ACOUSTIC WAVES IN SEMICONDUCTOR - PIEZOELECTRIC FILM STRUCTURES, Soviet Phys - Semiconductors v5 n1 p66 (Jly 71)

Gundlach, R., SURFACE WAVES PICK UP SPEED, Electronics v43 n24 p93 (23 Nov 70)

Gunston, W.T., SEARCHING FOR TOMORROW'S ELECTRONICS, Science J p 41 (May 70)

Gurevich, V.L., GROWTH OF FLUCTUATIONS IN AN UNSTABLE SYSTEM II, Soviet Phys - JETP v20 n4 p873 (Apr 65)

Gurevich, V.L., Kagan, V.D., RAYLEIGH LIGHT SCATTERING DURING SOUND INSTABILITY, Soviet Phys - JETP v20 n5 p1201 (May 65)

Gurevich, V.L., Kagan, V.D., Laikhtman, B.D., THEIRY OF THE INTERACTION OF GROWING SOUND FLUCTUATIONS IN SEMICONDUCTORS, Soviet Phys - JETP v27 n1 p102 (Jly 68)

Hagon, P.J., Pizzarello, F.A., Elkins, P.E., Coker, J.E., CHARACTERIZATION OF NEW MATERIALS FOR SURFACE-WAVE DEVICE APPLICATIONS (Abstr.), IEEE Trans Sonics & US vSU-17 n1 p59 (Jan 70)

Hagon, P.J., Sallee, G.F., Lakin, K.M., BISMUTH GERMANIUM OXIDE VERSUS LITHIUM NIOBATE IN THE SURFACE-WAVE AMPLIFIER. A THEORETICAL AND EXPERIMENTAL COMPARISON (Abstr.), IEEE Trans Sonics & US vSU-18 n1 p54 (Jan 71)

Hagon, P.J., Sallee, G.F., Lakin, K.M., U.H.F. ACOUSTIC SURFACE-WAVE DEVICES IN ALUMINUM NITRIDE FILMS ON SAPPHIRE (Abstr.) IEEE Trans Sonics & US vSU-18 n1 p66 (Jan 71)

Hagon, P.J., Austerman, S.B., ACOUSTIC WAVES IN BERYLLIUM OXIDE CRYSTAL, Appl Phys Lett v18 n3 p102 (1 Feb 71)

Haines, K.A., Hildebrand, B.P. SURFACE-DEFORMATION MEASUREMENT USING THE WAVEFRONT RECONSTRUCTION TECHNIQUE, Appl Opt v5 n4 p595 (Apr 66)

Hanebrekke, H., Ingebrigtsen, K.A., ACOUSTOELECTRIC AMPLIFICATION OF SURFACE WAVES IN STRUCTURE OF CADMIUM-SELENIDE FILM ON LITHIUM NIOBATE, Electron Lett v6 n16 p520 (6 Aug 70)

Hallermeier, R.J., Mayer, W.G., LIGHT DIFFRACTION BY ULTRASONIC SURFACE WAVES OF ARBITRARY STANDING-WAVE RATIO, Ac Soc Am J v47 n5(Pt.2) p1236 (May 70)

Hallermeier, R.J., Mayer, W.G., FREQUENCY SHIFTING OF A GAUSSIAN LIGHT BEAM BY ULTRASONIC SURFACE WAVES, J Appl Phys v41 n9 p3664 (Aug 70)

Hallermeier, R.J., Diachok, O.I., SIMPLE TECHNIQUE FOR EXCITING AND PROBING ELASTIC SURFACE WAVES, J Appl Phys v41 n12 p4763 (Nov 70)

Hallermeier, R.J., Mayer, W.G., SCATTERING OF A GAUSSIAN LIGHT BEAM BY REFLECTION FROM ATTENUATED SURFACE WAVES, IEEE Trans Sonics & US vSU-18 n3 p176 (Jly 71)

Hartmann, P., Dieulesaint, E., INTRINSIC COMPENSATION OF SIDELOBES IN A DISPERSIVE ACOUSTIC DELAY LINE, Electron Lett v5 n10 p219 (15 May 69)

Hartmann, P., Dieulesaint, E., ACOUSTIC-SURFACE-WAVE FILTERS, Electron Lett v5 n25 p657 (11 Dec 69)

Hartmann, C.S., Jones, W.S., WIDEBAND UNIDIRECTIONAL INTERDIGITAL SURFACE-WAVE TRANSDUCERS (Abstr.), IEEE Trans Sonics & US vSU-18 n1 p52 (Jan 71)

Hayakawa, H., Takada, S., Ishiguro, T., Mikoshiba, N., GENERATION OF SURFACE WAVES BY GUNN OSCILLATOR (Abstr.), IEEE Trans Sonics & US vSU-18 n1 p54 (Jan 71)

Hayakawa, H., Takada, S., Ishiguro, T., Mikoshiba, N., DETECTION OF SURFACE WAVES WITH GaAs BULK EFFECT DIODE, Appl Phys Lett v18 n1 p31 (1 Jan 71)

Hickernell, F.S., SHEAR MODE WEDGE TRANSDUCERS FOR SURFACE-WAVE EXCITATION IN THE V.H.F. REGION (Abstr.), IEEE Trans Sonics & US vSU-18 n1 p52 (Jan 71)

Hickernell, F.S., Shulda, G.F., Brewer, J.W., ZnO AND CdS OVERLAY SURFACE-WAVE TRANSDUCERS (Abstr.), IEEE Trans Sonics & US vSU-18 n1 p53 (jan 71)

Holland, M.G., Schulz, M.B., Tancrell, R.H., Barrett, H.H., PULSE EXPANSION AND COMPRESSION USING SURFACE ULTRASONIC WAVES, IEEE Int'l Conv Dig p94 (1969)

Humphryes, R.F., Ash, E.A., ACOUSTIC BULK-SURFACE-WAVE TRANSDUCER, Electron Lett v5 n9 p175 (1 May 69)

Hunsinger, G.J. Holshouser, D., SPATIAL MODULATION OF LIGHT USING SURFACE WAVES IN AN INTERFEROMETER, Appl Phys Lett v16 n7 p272 (1 Apr 70)

Hunsinger, B.J., OPTICAL GRATINGS FORMED BY SURFACE WAVES IN AN INTERFEROMETER (Abstr.), IEEE Trans Sonics & US vSU-18 n1 p56 (Jan 71)

Hunsinger, B.J., ANALYSIS OF AN INTERFEROMETER WITH ACOUSTIC SURFACE WAVES ON ONE MIRROR, Appl Opt v10 n2 p390 (Feb 71)

Hunsinger, B.J., Franck, A.R., PROGRAMMABLE SURFACE-WAVE TAPPED DELAY LINE, IEEE Trans Sonics & US vSU-18 n3 p152 (Jly 71)

Hutson, A.R., McFee, J.H., White, D.L., ULTRASONIC AMPLIFICATION IN CdS, Phys Rev Lett v7 n6 p237 (15 Sep 61)

Hutson, A.R., White, D.L., ELASTIC WAVE PROPAGATION IN PIEZOELECTRIC SEMI-CONDUCTORS, J Appl Phys v33 n1 p40 (Jan 62)

Ilukor, J., Jacobsen, E.H., GENERATION AND DETECTION OF COHERENT ELASTIC WAVES AT 114000 MHz/SEC, Science v153 p113 (2 Sep 66)

Ingebrigtsen, K.A., Tonning, A., NUMERICAL DATA FOR ACOUSTIC SURFACE WAVES IN ALPHA-QUARTZ AND CADMIUM SULFIDE, Appl Phys Lett v9 n1 p16 (1 Jly 66)

Ingebrigtsen, K.A., SURFACE WAVES IN PIEZOELECTRICS, J Appl Phys v40 n7 p2681 (Jne 69)

Ingebrigtsen, K.A., Tonning, A., ELASTIC SURFACE WAVES IN CRYSTALS, Phys Rev v184 n3 p942 (15 Aug 69)

Ingebrigtsen, K.A., Tonning, A., EXCITATION OF ELASTIC WAVES IN CRYSTALS, IEEE Trans Microwave T&T vMTT-17 n11 p827 (Nov 69)

Ingebrigtsen, K.A., LINEAR AND NONLINEAR ATTENUATION OF ACOUSTIC SURFACE WAVES IN A PIEZOELECTRIC COATED WITH A SEMICONDUCTOR FILM, J Appl Phys v41 n2 p454 (Feb 70)

Ippen, E.P., DIFFRACTION OF LIGHT BY SURFACE ACOUSTIC WAVES, IEEE Proc v40 n2 p248 (Feb 67)

Jaffe, H., Berlincourt, D.A., PIEZOELECTRIC TRANSDUCER MATERIALS, IEEE Proc v53 n10 p1372 (Oct 65)

Jaffe, H., PIEZOELECTRIC APPLICATIONS OF FERROELECTRICS, IEEE Trans Electron Dev vED-16 n6 p557 (jne 69)

Jones, W.R., Campbell, J.J., PROPAGATION OF SURFACE WAVES AT THE BOUNDARY BETWEEN A PIEZOELECTRIC CRYSTAL AND A FLUID MEDIUM (Abstr.), IEEE Trans Sonics & US vSU-17 n1 p58 (Jan 70)

Jones, W.S., Hartmann, C.S., Claiborne, L.T., EVALUATION OF DIGITALLY CODED ACOUSTIC SURFACE-WAVE MATCHED FILTERS, IEEE Trans Sonics & US vSU-18 n1 p21 (Jan 71)

Jones, W.S., Hartmann, C.S., Kempf, R.A., PRACTICAL LINEAR F.M. SURFACE-WAVE CORRELATORS (Abstr.), IEEE Trans Sonics & US vSU-18 n1 p60 (Jan 71)

Joseph, R.I., Schloemann, E., THEORY OF MAGNETOSTATIC MODES IN LONG, AXIALLY MEGNETIZED CYLINDERS, J Appl Phys v32 p1001 (Jne 61)

Joshi, S.G., White, R.M., DISPERSION OF SURFACE ELASTIC WAVES PRODUCED BY A CONDUCTING GRATING ON A PIEZOELECTRIC CRYSTAL, J Appl Phys v39 n13 p5819 (Dec 68)

Joshi, S.G., White, R.M., EXCITATION AND DETECTION OF SURFACE ELASTIC WAVES IN PIEZOELECTRIC CRYSTALS, Ac Soc Am J v46 n1 017 (Jly 69)

Joshi, S.G., Van den Heuvel, A.P., Serafin, R.J., A COMPOSITE SURFACE WAVE TRANSDUCER (Abstr.), IEEE Trans Sonics & US vSU-17 n1 p55 (Jan 70)

Kaganov, M.I., Sklovskaya, L.I., SURFACE WAVES IN A PIEZOELECTRIC, Soviet Phys -- Solid State v8 n12 p2789 (Jne 67)

Kahn, W.K., OMNIDIRECTIONAL TRANSDUCER, IBM Tech Discl Bull v12 n10 p1567 (Mch 70)

Katyl, R.H., EXTENDED RANGE SURFACE WAVE SCANNABLE LASER, IBM Tech Discl Bull v13 n7 p1825 (Dec 70)

Katyl, R.M., MICROSOUND F.M. DETECTOR, IBM Tech Discl Bull v13 n11 p3568 (Apr 71)

Kaufman, I., Foltz, J.W., SELF-SCANNED OPTICAL SENSOR USING ELASTIC SURFACE WAVES, IEEE Proc v57 n11 p2081 (Nov 69)

Kaufman, I., Crittenden, E.C., Oltman, H.G., DEVICES COMBINING PIEZOELECTRICITY AND PHOTOCONDUCTIVITY, (Abstr.), IEEE Trans Sonics & US vSU-17 n1 p56 (Jan 70)

Kazan, B., FIELD-EFFECT-CONTROLLED SURFACE WAVE DEVICE, IBM Tech Discl Bull v13 n4 p914 (Sep 70)

Keyes, R.W., VARIABLE ACOUSTIC DELAY LINE, IBM Tech Discl Bull v13 n8 p2435 (Jan 71)

Kharusi, M.S., Farnell, G.W., DIFFRACTION AND BEAM STEERING FOR SURFACE-WAVE COMB STRUCTURES ON ANISOTROPIC SUBSTRATES. IEEE Trans Sonics & US vSU-18 n1 p35 (Jan 71)

Kharusi, M.S., Farnell, G.W., THEORY AND EXPERIMENT IN THE DIFFRACTION OF QUASI-LONGITUDINAL, QUASI-SHEAR, AND RAYLEIGH WAVES (Abstr.), IEEE Trans Sonics & US vSU-18 n1 p47 (Jan 71)

Kim, Y.S., Smith, R.T., THERMAL EXPANSION OF LITHIUM TANTALATE AND LITHIUM NIOBATE SINGLE CRYSTALS, J Appl Phys v40 n11 p4637 (Oct 69)

Kino, G.S., GROWING SURFACE WAVES IN A SEMI CONDUCTOR IN THE PRESENCE OF A TRANSVERSE MAGNETIC FIELD, Appl Phys Lett v12 n9 p312 (1 May 68)

Knox, R.M., Van den Heuval, A.P., ELASTIC SURFACE WAVE CIRCUITS AT MICROWAVE FREQUENCIES, Proc 1970 National Electronics Conference

Koerber, G., UNCOUPLED PIEZOELECTRIC SURFACE-WAVE MODES, IEEE Trans Sonics & US vSU-18 n2 p73 (Apr 71)

Kolsky, H., "ELASTIC WAVES IN SOLIDS," Clarendon Press, Oxford (1953) and Dover Publications, Inc., New York, p9 (1963)

Korpel, A., Laub, L.J., Sievering, H.C., MEASUREMENT OF ACOUSTIC SURFACE WAVE PROPAGATION CHARACTERISTICS BY REFLECTED LIGHT, Appl Phys Lett v10 n10 p295 (15 May 67)

Korpel, A., ACOUSTIC IMAGING BY DIFFRACTED LIGHT I., TWO-DIMENSIONAL INTERACTION, IEEE Trand Sonics & US vSU-15 n3 p153 (Jly 68)

Kraetzig, E., Laub, L.J., Sievering, H.A., MEASUREMENT OF ACOUSTIC SURFACE WAVE PROPAGATION CHARACTERISTICS BY REFLECTED LIGHT, Appl Phys Lett v10 p295 (15 May 67)

Krairojananan, T., Redwood, M., PIEZOELECTRIC GENERATION AND DETECTION OF ULTRA-SONIC SURFACE WAVES BY INTERDIGITAL ELECTRODES: AN ELECTRICAL EQUIVALENT CIRCUIT, Electron Lett v5 n7 p134 (3 Apr 69)

Kraut, E.A., Tittmann, B.R., Graham, L.J., Lim, T.C., ACOUSTIC SURFACE WAVES ON METALLIZED AND UNMETALLIZED Bi_{12} Ge O_{20}, Appl Phys Lett v17 n7 p271 (1 Oct 70)

Kraut, E.A., Tittmann, B.R., Graham, L.J., Lim, T.C., ACOUSTIC SURFACE WAVES ON METALLIZED AND UNMETALLIZED Bi_{12} Ge O_{20}, (Abstr,), IEEE Trans Sonics & US vSU-18 n1 p66 (Jan 71)

Kravchenko, D.F., Tursunov, D.A., PHASE VELOCITIES OF SURFACE ELASTIC WAVES IN THE PLANE OF SYMMETRY OF THE ELASTIC PROPERTIES OF ALPHA-QUARTZ, Soviet Phys - Acoustics v16 n4 p529 (Apr/Jne 71)

Krischer, C., Bers, A., OBSERVED INTERACTION OF ACOUSTIC AND ELECTRON SURFACE WAVES IN A MAGNETIC FIELD, Appl Phys Lett v18 n8 p349 (15 Apr 71)

Krokstad, J., Svaasand, L.O., SCATTERING OF LIGHT BY ULTRASONIC SURFACE WAVES IN QUARTZ, Appl Phys Lett v11 n5 p155 (1 Sep 67)

Kuhn, L., INTEGRATED CIRCUIT RECIRCULATING GUIDED SURFACE WAVE ULTRASONIC STORE, IBM Tech Discl Bull v12 n10 p1569 (Mch 70)

Kuhn, L., Pennington, K.S., DYNAMIC GENERATION OF SWITCHABLE OPTICAL WAVEGUIDE COMPONENTS, IBM Tech Discl Bull v12 n11 p1944 (Apr 70)

Kuhn, L., Lean, E.G., Pennington, K.S., REAL-TIME ACOUSTO-OPTIC SURFACE WAVE DATA PROCESSOR, IBM Tech Discl Bull v13 n1 p105 (Jne 70)

Kuhn. L., MEASUREMENT ON SURFACE ACOUSTIC WAVES IN PHOTOCONDUCTING CADMIUM SULFIDE, IEEE Proc v58 n9 p1380 (Sep 70)

Kuhn. L., Dakss, M.L., Heidrich, P.F., Scott, B.A., DEFLECTION OF AN OPTICAL GUIDED WAVE BY A SURFACE ACOUSTIC WAVE, Appl Phys Lett v17 n6 p265 (15 Sep 70)

Lakin, K.M., ACOUSTOELECTRIC SURFACE-WAVE AMPLIFIERS, 1969 IEEE Int'l Conv Dig p92 (Mch 69)

Lakin, K.M., Collins, J.H., Hagon, P.J., 100 MHz SURFACE ACOUSTOELECTRIC AMPLIFIER EXHIBITING STABLE TERMINAL GAIN WITH dc DRIFT FIELD, IEEE Proc v57 n4 p740 (Apr 69)

Lakin, K.M., Shaw, H.J., SURFACE WAVE DELAY LINE AMPLIFIERS, IEEE Trans Microwave T&T VMTT-17 n11 p912 (Nov 69)

Lakin, K.M., SURFACE-WAVE VELOCITY PERTURBATION DUE TO AN ADJACENT ELECTRICALLY COUPLED SURFACE (Abstr.), IEEE Trans Sonics & US vSU-18 n1 p53 (Jan 71)

Lakin, K.M., PERTURBATION THEORY FOR ELECTROMAGNETIC COUPLING TO ELASTIC SURFACE WAVE ON PIEZOELECTRIC SUBSTRATES, J Appl Phys v42 n3 p899 (1 Mch 71)

Landau, L.D., Lifshitz, E.M., "THEORY OF ELASTICITY," Pergamon Press, London, p105 (1959)

Lapin, A.D., PROPAGATION OF SURFACE WAVES OVER CURVED SURFACES, Soviet Phys - Acoustics v15 n2 p201 (Oct/Dec 69)

Lapin, A.D., SCATTERING OF SURFACE WAVES PROPAGATING OVER AN UNEVEN LIQUID-SOLID INTERFACE, Soviet Phys - Acoustics v15 n3 p336 (Jan/Mch 70)

Lardat, C., Tournois, P., 100MHz BANDWIDTH LOVE WAVE DISPERSIVE DELAY LINES, IEEE Proc v58 n7 p1150 (Jly 70)

Lardat, C., Maerfeld, C., Tournois, P., THEORY AND PERFORMANCE OF ACOUSTIC DISPERSIVE SURFACE WAVE DELAY LINES, IEEE Proc v59 n3 p355 (Mch 71)

Lawes, G., NEW WAVE IN ACOUSTICS, New Scientist p82 (9 Oct 69)

Lawson, R.L., Hunsinger, B.J., Cho, F.Y., VISUALIZATION OF ACOUSTIC SURFACE WAVES IN AN EXPERIMENTAL FABRY-PEROT INTERFEROMETER, Appl Opt v9 n12 p2805 (Dec 70)

Lean, E.G.H., Powell, C.G., Kuhn, L., ACOUSTIC SURFACE WAVE MIXING ON ALPHA-QUARTZ, Appl Phys Lett v15 n1 p10 (1 Jly 69)

Lean, E.G.H., Broers, A.N., 2.5 GHz INTERDIGITAL SURFACE-WAVE TRANSDUCERS FABRICATED BY AN ELECTRON BEAM (Abstr.), IEEE Trans Sonics & US vSU-17 n1 p69 (Jan 70)

Lean, E.G.H., Tseng, C.C., Powell, C.G., OPTICAL PROBING OF ACOUSTIC SURFACE-WAVE HARMONIC GENERATION, Appl Phys Lett v16 n1 p32 (1 Jan 70)

Lean, E.G.H., Tseng, C.C., NONLINEAR EFFECTS IN SURFACE ACOUSTIC WAVES, (Abstr.) IEEE Trans Sonics & US vSU-17 n1 p53 (Jan 70)

Lean, E.G.H., Broers, A.N., S-BAND SURFACE ACOUSTIC DELAY LINE, 1970 IEEE Int'l Solid-State Circuits Conf p 130

Lean, E.G.H., GENERATION OF WIDE BAND ACOUSTICAL WAVES, IBM Tech Discl Bull v13 n1 p269 (Jne 70)

Lean, E.G.H., Yu, H.N., LOVE WAVE DEVICES, IBM Tech Discl Bull v13 12 p417 (Jly 70)

Lean, E.G.H., Tseng, C.C., NONLINEAR EFFECTS IN SURFACE ACOUSTIC WAVES, J Appl Phys v41 n10 p3912 (Sep 70)

Lean, E.G.H., Powell, C.G., OPTICAL PROBING OF SURFACE ACOUSTIC WAVES, IEEE Proc v58 n12 p3939 (Dec 70)

Lean, E.G.H., Pole, R.V., GALLIUM ARSENIDE GUNN OSCILLATOR TO EXCITE SURFACE ACOUSTIC WAVES, IBM Tech Discl Bull v13 n8 p2411 (Jan 71)

Lee, R.E., White, R.M., EXCITATION OF SURFACE ELASTIC WAVES BY TRANSIENT SURFACE HEATING, Appl Phys Lett v12 n1 p12 (1 Jan 68)

Lee, R.E., White, R.M., ELASTIC WAVE GENERATION BY A GUNN EFFECT OSCILLATOR COUPLED TO A PIEZOELECTRIC, Appl Phys Lett v16 n9 p343 (1 May 70)

Lee, R.E., White, R.M., BULK AND SURFACE-WAVE TRANSDUCTION UTILIZING A GUNN EFFECT OSCILLATOR ON A PIEZOELECTRIC (Abstr.), IEEE Trans Sonics & US vSU-18 n1 p54 (Jan 71)

Leedom, D.A., Krimholz, R., Matthaei, G.L., EQUIVALENT CIRCUIT FOR TRANSDUCERS HAVING ARBITRARY EVEN - OR ODD-SYMMETRY PIEZOELECTRIC EXCITATION (Abstr.), IEEE Trans Sonics & US vSU-18 n1 p61 (Jan 71)

Lewis, M.F., Patterson, E., NOVEL HELICAL-PATH SURFACE WAVE DELAY LINE, Appl Phys Lett v18 n4 p143 (15 Feb 71)

Lewis, M.F., Patterson, E., SOME PROPERTIES OF ACOUSTIC SURFACE WAVES ON X-CUT QUARTZ, Ac Soc Am J v49 n5(Pt.2) p1667 (May 71)

Lim, T.C., Farnell, G.W., SEARCH FOR FORBIDDEN DIRECTIONS OF ELASTIC SURFACE-WAVE PROPAGATION IN ANISOTROPIC CRYSTALS, J Appl Phys v39 n9 p4319 (Aug 68)

Lim, T.C., Farnell, G.W., CHARACTER OF PSEUDOSURFACE WAVES ON ANISOTROPIC CRYSTALS, Ac Soc Am J v45 n4 p845 (Apr 69)

Lim, T.C., Kraut, E.A., Tittmann, B.R., DISPERSIVE PROPERTIES OF THE FUNDAMENTAL RAYLEIGH WAVE MODE (M_{18}) IN A HETERO EPITAXIAL ZINC OXIDE FILM ON A SAPPHIRE SUBSTRATE, Appl Phys Lett v17 n1 p34 (1 Jly 70)

Lim, T.C., Kraut, E.A., EFFECT OF MASS LOADING ON THE PROPAGATION OF ACOUSTIC SURFACE WAVES ON $LiNbO_3$ AND QUARTZ (Abstr.), IEEE Trans Sonics & US vSU-18 n1 p67 (Jan 71)

Loftus, D.S., ELASTIC SURFACE WAVES ON LAYERED ANISOTROPIC CRYSTALS, Appl Phys Lett v13 n9 p323 (1 Nov 68)

Lopen, P.O., SECOND-HARMONIC GENERATION IN AN ELASTIC SURFACE WAVE IN ALPHA-QUARTZ, J Appl Phys v39 n12 p5400 (Nov 68)

Love, A.E.H.,"A TREATISE ON THE MATHEMATICAL THEORY OF ELASTOCITY," Dover Publications, Inc., New York, p99 (1944)

Ludloff, H., ON ULTRASONIC SURFACE WAVES AND THEIR OPTICAL DETECTION, Zeitschr. Tech Phyz v17 p518 (1936) in German

Ludloff, H., ON ULTRASONIC SURFACE WAVES, Phyz Zeitschr v37 p524 (1936) in Germ.

Luukkala, M., Kino, G.S., CONVOLUTION AND TIME INVERSION USING PARAMETRIC INTER-ACTIONS OF ACOUSTIC SURFACE WAVES, Appl Phys Lett v18 n9 p393 (1 May 71)

Lyamshev, L.M., SURFACE WAVE AMPLIFICATION, Soviet Phys - Acoustics v16 n2 p270 (Oct/Dec 70)

Maerfeld, C., Lardat, C., ONDES DE LOVE AT DE RAYLEIGH DANS LES MILIEUX PIEZOELECTRIQUES STRATIFIES, Compt Rend v270(Ser.B) n19 p1187 (11 May 70)

Maerfeld, C., AMPLIFICATION OF BLEUSTEIN WAVES IN PIEZOELECTRIC SEMICONDUCTORS (Abstr.), IEEE Trans Sonics & US vSU-18 n1 p53 (Jan 71)

Maerfeld, C., Gires, F., Tournois, P., BLEUSTEIN-GULYAEV SURFACE WAVE AMPLIFICATION IN CdS, Appl Phys Lett v18 n7 p269 (1 Apr 71)

Maradudin, A.A., Mills, D.L., CALCULATION OF THE ANHARMONIC DAMPING OF RAYLEIGH SURFACE MODES, Phys Rev v173 n3 p881 (15 Sep 68)

Mason, W.P., "ELECTRO-MECHANICAL TRANSDUCERS AND WAVE FILTERS," 2nd ed., Van Nostrand, Princeton, New Jersey, p201-209, 399-409 (1948)

Mason, I.A., ACOUSTOELECTRIC INTERACTIONS IN CADMIUM-SULPHIDE THIN FILMS, Electron Lett v5 n21 p519 (16 Oct 69)

Matthews, H., NONRECIPROCAL SURFACE WAVE DEVICES, N.E.R.E.M. Record v10 p176 (1968)

Matthews, H., Van de Vaart, H., OBSERVATION OF LOVE WAVE PROPAGATION AT U.H.F. FREQUENCIES, Appl Phys Lett v14 n5 p171 (1 Mch 69)

Matthews, H., Van de Vaart, H., MAGNETO-ELASTIC LOVE WAVES, Appl Phys Lett v15 n11 p373 (1 Dec 69)

Mayer, W.G., Lamers, G.B., Auth, D.C., INTERACTION OF LIGHT AND ULTRASONIC SURFACE WAVES, Ac Soc Am J v42 n6 p1255 (Dec 67)

McBride, S.L., ON THE INTERACTION OF ACOUSTIC SURFACE WAVES WITH THERMAL PHONONS AT LOW TEMPERATURES, Canadian J Phys v47 n10 p1143 (15 May 69)

McFee, J.H., "PHYSICAL ACOUSTICS," W.P. Mason, ed., vol IVA, New York: Academic Press, p1-45 (1966)

Meeker, T.R., DETECTION OF ELASTIC WAVES BY A PIEZOELECTRIC TRANSDUCER WITH A GRATING ELECTRODE - APPLICATION TO ULTRASONIC DIFFRACTION DELAY LINES AND SURFACE-WAVE DEVICES (Abstr.), IEEE Trans Sonics & US vSU-18 n1 p59 (Jan 71)

Milsom, R.F., Redwood, M., PIEZOELECTRIC COUPLING COEFFICIENT OF BLEUSTEIN-GULYAEV WAVES, Electron Lett v6 n21 1665 (15 Oct 70)

Mitchell, R.F., Willis, W., Redwood, M., ELECTRODE INTERACTIONS IN ACOUSTIC SURFACE-WAVE TRANSDUCERS, Electron Lett v5 n19 p456 (18 Sep 69)

Mizushima, Y., Sudo, T., SURFACE-WAVE AMPLIFICATION BETWEEN PARALLEL SEMI-CONDUCTORS, IEEE Trans Electron Dev vED-17 n7 (Jly 70)

Montgomery, R.M., Young, E.H.Jr., LIGHT DIFFRACTION BY AN ACOUSTIC SURFACE WAVE IN FUSED QUARTZ, J Appl Phys v42 n7 p2585 (Jne 71)

Morgan, S.P., Ash, E.A., ACOUSTIC SURFACE WAVE DISPERSIVE DELAY LINES, Proc of IEE (London) v116 n7 p1125 (Jly 69)

Motamedi, M.E., Epstein, M., WIDE-BAND TRANSDUCER FOR SURFACE ELASTIC WAVES, IEEE Proc v58 n11 p1870 (Nov 70)

Myshkin, V.G., DIFFRACTION OF A SCALAR SURFACE WAVE AT OBLIQUE INCIDENCE ON A BOUNDARY BETWEEN IMPEDANCE HALF-PLANES, Soviet Phys - Acoustics v12 n3 p300 (Jan/Mch 67)

Nassenstein, H., SUPERRESOLUTION BY DIFFRACTION OF SUBWAVES, Opt Commun v2 n5 p231 (Oct 70)

Ogg, N.R., HUYGENS PRINCIPLE FOR ANISOTROPIC CRYSTAL SURFACE (Abstr.), IEEE Trans Sonics & US vSU-18 n1 p47 (Jan 71)

Oliner, A.A., MICROWAVE NETWORK METHODS FOR GUIDED ELASTIC WAVES, IEEE Trans Microwave T&T vMTT-17 n11 p812 (Nov 69)

Olson, F.A., Kirchner, E.K., Mehta, K.B., Peternell, F.J., Morley, B.C., PROPAGATION OF MAGNETOSTATIC SURFACE WAVES IN Y.I.G. RODS, J Appl Phys v38 n3 p1218 (Mch 67)

Owens, J.K., Sallee, G.F., SURFACE ACOUSTIC WAVE LOGIC GATES, IEEE Proc v59 n2 p308 (Feb 71)

Parekh, J.P., MAGNETOELASTIC SURFACE WAVE IN FERRITES, Electronics Lett v5 n14 p322 (10 Jly 69)

Parekh, J.P., PROPAGATION CHARACTERISTICS OF MAGNETOELASTIC SURFACE WAVE, Electron Lett v5 n21 p540 (16 Oct 69)

Parekh, J.P., INFLUENCE OF SURFACE METALLIZATION ON MAGNETOELASTIC SURFACE WAVES, Electron Lett v6 n2 p47 (22 Jan 70)

Parekh, J.P., Bertoni, H.L., MAGNETOELASTIC RAYLEIGH-TYPE SURFACE WAVE (Abstr.), IEEE Trans Sonics & US vSU-18 n1 p51 (Jan 71)

Pedinoff, M.E., Waldner, M., REFRACTION AND REFLECTION OF ULTRASONIC SURFACE WAVES BY THIN-GOLD-FILM LAYERS ON LITHIUM NIOBATE SUBSTRATES, Electron Lett v6 n17 p533 (20 Aug 70)

Pedinoff, M.E., Waldner, M., Jones, W.R., OPTICAL STUDIES OF REFRACTION AND REFLECTION OF ULTRASONIC SURFACE WAVES BY THIN GOLD FILM LAYERS ON LITHIUM NIOBATE SUBSTRATES (Abstr.), IEEE Trans Sonics & US vSU-18 n1 p66 (Jan 71)

Peercy, P.S., Jones, E.D., Bushnell, J.C., Gobeli, G.W., ULTRAFAST RISE TIME LASER-INDUCED STRESS WAVES, Appl Phys Lett v16 n3 p120 (1 Feb 70)

Pratt, R.G., Lim, T.C., ACOUSTIC SURFACE WAVES ON SILICON, Appl Phys Lett v15 n12 p403 (15 Dec 69)

Press, F., Healy, J., ABSORPTION OF RAYLEIGH WAVES IN LOW-LOSS MEDIA, J Appl Phys v28 p1323 (Nov 57)

Pulliam, G.R., EPITAXIAL MATERIALS FOR MICROSOUND APPLICATIONS, N.E.R.E.M. Rec p174 (Nov 68)

Pulliam, G.R., Collins, J.H., Hagon, P.J., EVALUATION OF NEW SINGLE CRYSTAL PIEZO-ELECTRIC MATERIALS FOR SURFACE ACOUSTIC WAVE APPLICATIONS, (Abstr.), IEEE Trans Sonics & US vSU-18 n1 p49 (Jan 71)

Quate, C.F., Wilkinson, C.D., Winslow, D.K., INTERACTION OF LIGHT AND MICROWAVE SOUND, IEEE Proc v53 n10 p1604 (Oct 65)

Quate, C.F., AMPLIFICATION OF ACOUSTIC SURFACE WAVES WITH TWO SETS OF CARRIERS, Electron Lett v5 n14 p317 (10 Jly 69)

Rabinowicz, E., Kuhn, L., SURFACE ACOUSTIC WAVE MONITOR FOR DETECTION OF WEAR IN DISK MEMORIES, IBM Tech Discl Bull v13 n11 p3236 (Apr 71)

Rabinowicz, E., SPLIT INTERDIGITAL TRANSDUCERS, IBM Tech Discl Bull v13 n11 p3257 (Apr 71)

Rayleigh, Lord, ON WAVES PROPAGATED ALONG THE PLANE SURFACE OF AN ELASTIC SOLID, Proc London Math Soc, vXVII p4 (12 Nov 1885)

Redwood, M., Milsom, R.F., PIEZOELECTRIC COUPLING COEFFICIENT OF INTERDIGITAL RAYLEIGH-WAVE TRANSDUCERS, Electron Lett v6 n14 p437 (9 Jly 70)

Reeder, T.M., Westbrook, E.M., Winslow, D.K., VISUALIZATION OF SURFACE ACOUSTIC WAVES, Electron Lett v6 n2 p30 (22 Jan 70)

Reeder, T.M., Shaw, H.J., Westbrook, E.M., NOISE AND SATURATION CHARACTERISTICS OF THE SEPARATED MEDIUM RAYLEIGH WAVE AMPLIFIER (Abstr.), IEEE Trans Sonics & US vSU-18 n1 p54 (Jan 71)

Richardson, B.A., Kino, G.S., PROBING OF ELASTIC SURFACE WAVES IN PIEZOELECTRIC MEDIA, Appl Phys Lett v16 n2 p82 (15 Jan 70)

Rodrigne, G.P., MICROWAVE SOLID-STATE DELAY LINES, IEEE Proc v53 n10 p1428 (Oct 65)

Rollins, F.R., Lim, L.T.C., Farnell, G.W., ULTRASONIC REFLECTIVITY AND SURFACE WAVE PHENOMENA ON SURFACES OF COPPER SINGLE CRYSTALS, Appl Phys Lett v12 n7 p236 (1 Apr 68)

Rosenfeld, R.C., BLEUSTEIN WAVES ON CYLINDRICAL SURFACE (Abstr.), IEEE Trans Sonics & US vSU-18 n1 p48 (Jan 71)

Ross, J.D., Kapuscienski, S.J., Daniels, K.B., VARIABLE DELAY LINE USING ULTRASONIC SURFACE WAVES, 1958 IRE National Conv Rec v6 p118

Route, R.K., Kino, G.S., SOUND WAVE AMPLIFICATION IN INDIUM ANTIMONIDE, Appl Phys Lett v14 n3 p97 (1 Feb 69)

Rulf, B., RAYLEIGH WAVES ON CURVED SURFACES, Ac Soc Am J v45 n2 p493 (Feb 69)

Sabine, P.V.H., RAYLEIGH-WAVE PROPAGATION ON A PERIODICALLY ROUGHENED SURFACE, Electron Lett v6 n6 p149 (19 Mch 71)

Sadagopan, V., Kuhn, L., ACOUSTO-OPTIC DEFLECTOR, IBM Tech Discl Bull v13 n1 p99 (Jne 70)

Salzmann, E., Plieninger, T., Dransfeld, K., ATTENUATION OF ELASTIC SURFACE WAVES IN QUARTZ AT FREQUENCIES OF 316 MHz and 1047 MHz, Appl Phys Lett v13 n1 p14 (1 Jly 68)

Salzmann, E., D. Weisman, OPTICAL DETECTION OF RAYLEIGH WAVES, J Appl Phys v40 n8 p3408 (Jly 69)

Schloemann, E., AMPLIFICATION OF MAGNETOSTATIC SURFACE WAVES BY INTERACTION WITH DRIFTING CHARGE CARRIERS IN CROSSED ELECTRIC AND MAGNETIC FIELDS, J Appl Phys v40 n3 p1422 (Mch 69)

Schmidt, R.V., Voltmer, F.W., PIEZOELECTRIC ELASTIC SURFACE WAVES IN ANISOTROPIC LAYERED MEDIA, IEEE Trans Microwave T&T vMTT-17 n11 p920 (Nov 69)

Schmidt, R.V., OPTICAL PROBING OF BULK WAVES PRESENT IN ACOUSTIC SURFACE WAVE DELAY LINES, Appl Phys Lett v17 n9 p369 (1 Nov 70)

Schnitzler, P., ELASTIC SURFACE WAVES IN LAYERED SYSTEM OF CADMIUM SULFIDE ON GERMANIUM, Appl Phys Lett v11 n9 p273 (1 Nov 67)

Schnitzler, P., Bergstein, L., Strauss, L., ELASTIC SURFACE WAVES: THIN FILM TRANSDUCERS AND LAYERED-SYSTEM DISPERSION, IEEE Trans Sonics & US vSU-17 n3 p185 (Jly 70)

Schulz, M.B., Matsinger, B.J., Holland, M.G., TEMPERAURE DEPENDENCE OF THE SURFACE ACOUSTIC WAVE VELOCITY IN QUARTZ, 1969 IEEE Ultrasonic Symp (Abstr.), IEEE Trans Sonics & US vSU-17 n1 p59 (Jan 70)

Schulz, M.B., Matsinger, B.J., Holland, M.G., TEMPERATURE DEPENDENCE OF SURFACE ACOUSTIC WAVES VELOCITY ON ALPHA-QUARTZ, J Appl Phys v41 n7 p2755 (Jne 70)

Schultz, M.B., Holland, M.G., SURFACE ACOUSTIC WAVE DELAY LINES WITH SMALL TEMPERATURE COEFFICIENT, IEEE Proc v58 n9 p1361 (Sep 70)

Schultz, M.B., Meyer, P.C., RAYLEIGH WAVE PROPAGATION ON CURVED SURFACES OF CRYSTALS (Abstr.), IEEE Trans Sonics & US vSU-18 n1 p48 (Jan 71)

Schultz, M.B., Holland, M.G., TEMPERATURE DEPENDENCE OF SURFACE ACOUSTIC WAVE VELOCITY ON $LiTaO_3$ (Abstr.), IEEE Trans Sonics & US vSU-18 n1 p66 (Jan 71)

Serafin, R.J., Fryberger, D., Epstein, M., MEASUREMENT OF STRAIN AND VELOCITY OF ULTRASONIC SURFACE WAVES, Rev Sci Instr v38 n10 p1395 (Oct 67)

Seshadri, S.R., SURFACE MAGNETOSTATIC MODES OF A FERRITE SLAB, IEEE Proc v58 n3 p506 (Mch 70)

Shaw, H.J., PARAMETRIC ACOUSTIC INTERACTIONS: A NEW APPROACH TO SIGNAL PROCESSING (Abstr.), IEEE Trans Sonics & US vSU-18 n1 p51 (Jan 71)

Sittig, E.C., Coquin, G.A., FILTERS AND DISPERSIVE DELAY LINES USING REPETITIVELY MISMATCHED ULTRASONIC TRANSMISSION LINES, IEEE Trans Sonics & US vSU-15 n2 p111 (Apr 68)

Skeie, H., ELECTRICAL AND MECHANICAL LOADING OF A PIEZOELECTRIC SURFACE SUPPORTING SURFACE WAVES (Abstr.), IEEE Trans Sonics & US vSU-17 n1 p55 (Jan 70)

Skeie, H., ELECTRICAL AND MECHANICAL LOADING OF A PIEZOELECTRIC SURFACE SUPPORTING SURFACE WAVES, Ac Soc Am J v48 n5(Pt.2) p1098 (Nov 70)

Slobodnik, A.J.,Jr., Carr, P.H., MICROWAVE FREQUENCY ACOUSTIC SURFACE WAVES, (Abstr.), IEEE Trans Sonics & US vSU-16 n1 p23 (Jan 69)

Slobodnik, A.J.,Jr., MICROWAVE FREQUENCY ACOUSTIC WAVE PROPAGATION LOSSES IN

$LiNbO_3$, Appl Phys Lett v14 n3 p94 (1 Feb 69)

Slobodnik, A.J., Jr., NONLINEAR EFFECTS IN MICROWAVE ACOUSTIC SURFACE WAVE DELAY LINES, (Abstr.), IEEE Trans Sonics & US vSU-17 n1 p53 (Jan 70)

Slobodnik, A.J.,Jr., MICROWAVE ACOUSTIC SURFACE WAVE INVESTIGATION USING LIGHT DIFFRACTION, IEEE Proc v58 n3 p488 (Mch 70)

Slobodnik, A.J.,Jr., Conway, E.D., NEW HIGH-FREQUENCY HIGH-COUPLING LOW-BEAM-STEARING CUT FOR ACOUSTIC SURFACE WAVES ON $LiNbO_3$, Electron Lett v6 n6 p171 (19 Mch 70)

Slobodnik, A.J.,Jr., NONLINEAR EFFECTS IN MICROWAVE ACOUSTICS $LiNbO_3$ SURFACE-WAVE DELAY LINES, Ac Soc Am J v48 n1(Pt.2) p203 (Jly 70)

Slobodnik, A.J.,Jr., Carr, P.H., Budreau, A.J., MICROWAVE FREQUENCY ACOUSTIC SURFACE WAVE LOSS MECHANISMS IN $LiNbO_3$, J Appl Phys v41 n11 p4380 (Oct 70)

Slobodnik, A.J.,Jr., SURFACE-WAVE ATTENUATION AT MICROWAVE FREQUENCIES: PART I - ROOM TEMPERATURE (Abstr.), IEEE Trans Sonics &US vSU-18 n1 p50 (Jan 71)

Smith, A.B., Kestigian, M., Kedzie, R.W., Grace, M.I., SHEAR-WAVE ATTENUATION IN LITHIUM NIOBATE, J Appl Phys v38 n12 p4928 (Nov 67)

Smith, A.B., Kedzie, R.W., McMahon, D.H., Kestigian, M., MICROWAVE ULTRASONIC ATTENUATION IN $LiTaO_3$, J Appl Phys v40 n7 p2687 (Jne 69)

Smith, A.B., Damon, R.W., A BIBLIOGRAPHY OF MICROWAVE ULTRASONICS, IEEE Trans Sonics & US vSU-17 n2 p86 (Apr 70)

Smith, A.B., MICROWAVE MAGNETOELASTIC DEVICES, Proc 20th Electronic Components Conf., p480 (13-15 May 70)

Smith, H.I., THE FABRICATION OF MICROSOUND COMPONENTS, 1969 IEEE Int'l Conv Dig p90 (Mch 69)

Smith, H.I., METHOD SOR FABRICATING HIGH-FREQUENCY SURFACE WAVE TRANSDUCERS, Rev Sci Instr v40 n5 p729 (May 69)

Smith, H.I., Bachner, F.J., HIGH YIELD TECHNIQUE FOR THE FABRICATION OF SURFACE WAVE DEVICES (Abstr.), IEEE Trans Sonics & US vSU-18 n1 p67 (Jan 71)

Smith, W.R., Gerard, H.M., Collins, J.H., Reeder, T.M., Shaw, H.J., ANALYSIS OF INTERDIGITAL SURFACE WAVE TRANSDUCERS BY USE OF AN EQUIVALENT CIRCUIT MODEL, IEEE Trans Microwave T&T vMTT-17 n11 p856 (Nov 69)

Smith, W.R., Gerard, H.M., Collins, J.H., Reeder, T.M., Shaw, H.J., DESIGN OF SURFACE WAVE DELAY LINES WITH INTERDIGITAL TRANSDUCERS, IEEE Trans Microwave T&T vMTT-17 n11 p865 (Nov 69)

Smith, W.R., Reeder, T.M., Collins, J.H., Shaw, H.J., DISPERSIVE RAYLEIGH WAVE DELAY LINE UTILIZING GOLD ON LITHIUM NIOBATE, IEEE Trans Microwave T&T vMTT-17 n11 p1043 (Nov 69)

Soffer, B.H., Close, D.H., Pedinoff, M.E., AN OPTICAL IMAGING METHOD FOR DIRECT OBSERVATION AND STUDY OF ACOUSTIC SURFACE WAVES, Appl Phys Lett v15 n10 p339 (15 Nov 69)

Solie, L.P., PIEZOELECTRIC ACOUSTIC SURFACE WAVES FOR A FILM ON SUBSTRATE, Appl Phys Lett v18 n4 p111 (15 Feb 71)

Sparks, M., MAGNETOSTATIC SURFACE MODES OF Y.I.G. SLAB, Electronics Letters v5 n24 p618 (27 Nov 69)

Spears, D.L., Burke, B.E., LOW POWER DENSITY GaAs/$LiNbO_3$ SURFACE-WAVE AMPLIFIER, (Abstr.), IEEE Trans Sonics & US vSU-18 n1 p54 (Jan 71)

Spector, H.N., AMPLIFICATION OF ACOUSTIC WAVES THROUGH INTERACTION WITH CONDUCTION ELECTRONS, Phys Rev v127 n4 p1084 (15 Aug 62)

Speiser, J.M., Whitehouse, H.J., NEW SURFACE-WAVE TRANSDUCER CODINGS (Abstr.), IEEE Trans Sonics & US vSU-18 n1 p61 (Jan 71)

Squire, W.D., Whitehouse, H.J., Alsup, J.M., LINEAR SIGNAL PROCESSING AND ULTRASONIC TRANSVERSAL FILTERS, IEEE Trans Microwave T&T vMTT-17 n11 p1020 (Nov 69)

Stern, E., MICROSOUND COMPONENTS, CIRCUITS, AND APPLICATIONS, IEEE Trans Microwave T&T vMTT-17 n11 p835 (Nov 69)

Stern. E., AN ASSASSMENT OF MICROSOUND TECHNOLOGY (Abstr.), IEEE Trans Sonics & US vSU-18 n1 p44 (Jan 71)

Stoneley, R., ELASTIC WAVES AT THE SURFACE OF SEPARATION OF TWO SOLIDS, Proc Roy Soc (London) v106A n738A p416 (1 Oct 24)

Stoneley, R., THS PROPAGATION OF SURFACE ELASTIC WAVES IN A CUBIC CRYSTAL, Proc Roy Soc (London) v232A n1191 p447 (22 Nov 55)

Szabo, T.L., Slobodnik, A.J.,Jr., EFFECT OF DIFFRACTION ON THE DESIGN OF MICROWAVE ACOUSTIC SURFACE WAVE DEVICES (Abstr.), IEEE Trans Sonics & US vSU-18 n1 p47 (Jan 71)

Svaasand, L.O., INTERACTION BETWEEN ELASTIC SURFACE WAVES IN PIEZOELECTRIC MATERIALS, Appl Phys Lett v15 n9 p300 (1 Nov 69)

Synge, J.L., ELASTIC WAVES IN ANISOTROPIC MEDIA, J Math & Phys v35 n4 p323 (Jan 57)

Takayanagi, K., Yamanouchi, K., Shibayoma, K., PIEZOELECTRIC LEAKY SURFACE WAVE IN $LiNbO_3$ Appl Phys Lett v17 n5 p225 (1 Sep 70)

Tancrell, R.H., Schultz, M.B., Barrett, H.H., Davis, L.,Jr., Holland, M.G., DISPERSIVE DELAY LINES USING ULTRASONIC SURFACE WAVES, IEEE Proc v57 n6 p1211 (Jne 69)

Tancrell, R.H., Holland, M.G., THE USE OF SURFACE ULTRASONIC WAVES FOR GENERALIZED FILTERS OF U.H.F., IEEE Trans Sonics & US vSU-17 n1 p60 (Jan 70)

Tancrell, R.H., SURFACE-WAVE FILTERS (Abstr.), IEEE Trans Sonics & US vSU-18 n1 p60 (Jan 71)

Tancrell, R.H., ACOUSTIC SURFACE WAVE FILTERS, IEEE Proc v59 n3 p393 (Mch 71)

Tannenwald, P.E., MICROWAVE ULTRASONICS, Microwave J v6 p61 (1963)

Tao, T.F., Tully, J.W., Schott, F.W., DYNAMIC MODE SURFACE WAVES ON MAGNETIZED Y.I.G. AND FERRITE RODS, Appl Phys Lett v14 n3 p106 (1 Feb 69)

Tao, T.F., Tully, J.W., NONMAGNETIC VOLUME AND SURFACE-WAVE MODES ON GYROMAGNETIC Y.I.G. ROD WAVEGUIDES, Appl Phys Lett v16 n5 p194 (1 Mch 70)

Tao, T.Y., DYNAMIC CONTROL OF LASER DEFLECTION BY ACOUSTIC SURFACE WAVES, IBM Tech Discl Bull v13 n6 p1603 (Nov 70)

Tehon, S.W., Wanuga, S., MICROWAVE ACOUSTICS, IEEE Proc v52 p1113 (Oct 64)

Teodori, E., SURFACE-WAVE MATCHED DELAY LINES FOR PULSE COMPRESSION, Electron Lett v5 n15 p334 (24 Jly 69)

Thomann, H., COUPLING MEASUREMENTS OF SURFACE-WAVE INTERDIGITAL TRANSDUCERS ON PIEZOELECTRIC CERAMICS, Electron Lett v5 n25 p652 (11 Dec 69)

Thomann, H., NEW TYPE OF RAYLEIGH-WAVE TRANSDUCER ON PIEZOELECTRIC CERAMICS, Electron Lett v6 n22 p710 (29 Oct 70)

Tiersten, H.F., ELASTIC SURFACE WAVES GUIDED BY THIN FILMS, J Appl Phys v40 n2 p770 (Feb 69)

Tiersten, H.F., ON MAGNETIC EXCHANGE BOUNDARY CONDITIONS IN MAGNETOELASTICITY (Abstr.), IEEE Trans Sonics & US vSU-17 n1 p69 (Jan 70)

Tiersten, H.F., ON THE NONLINEAR EQUATIONS OF ELECTROELASTICITY (Abstr.), IEEE Trans Sonics & US vSU-18 n1 p51 (Jan '71)

Timoshenko, S., "THEORY OF ELASTICITY," McGraw-Hill Book Co., Inc., New York, p400 (1934)

Tittmann, B.R., Kraut, E.A., Lim, T.C., DIRECT MEASUREMENT OF SURFACE WAVE VELOCITY AND TIME DELAY - TEMPERATURE VARIATION USING A PULSE TECHNIQUE, IEEE Proc v59 n1 p86 (Jan 71)

Tittmann, B.R., ACCURATE DETERMINATIONS OF ACOUSTIC SURFACE-WAVE VELOCITY, ELECTROMECHANICAL COUPLING, AND TEMPERATURE-DELAY COEFFICIENTS (Abstr.), IEEE Trans Sonics & _S vSU-18 n1 p66 (Jan 71)

Tournois, P., Lardat, C., LOVE WAVE - DISPERSIVE DELAY LINES FOR WIDEBAND PULSE COMPRESSION, IEEE Trans Sonics & US vSU-16 n3 p107 (Jly 69)

Tournois, P., Bertheas, J., USE OF DISPERSIVE DELAY LINES FOR SIGNAL PROCESSING IN UNDERWATER ACOUSTICS, Ac Soc Am J v46 (Pt.1) p517 (Sep 69)

Tseng, C.C., White, R.M., PROPAGATION OF PIEZOELECTRIC AND ELASTIC SURFACE WAVES ON THE BASAL PLANE OF HEXAGONAL PIEZOELECTRIC CRYSTALS, J Appl Phys v38 n11 p4274 (Oct 67)

Tseng, C.C., ELASTIC SURFACE WAVES ON FREE SURFACE AND METALLIZED SURFACE OF CdS, ZnO, AND PZT-4, J Appl Phys v38 n11 p4280 (Oct 67)

Tseng, C.C., FREQUENCY RESPONSE OF AN INTERDIGITAL TRANSDUCER FOR EXCITATION OF SURFACE ELASTIC WAVES, IEEE Trans Electron Dev vED-15 n8 p586 (Aug 68)

Tseng, C.C., PIEZOELECTRIC SURFACE WAVES IN CUBIC AND ORTHORHOMBIC CRYSTALS, Appl Phys Lett v16 n6 p253 (15 Mch 70)

Tseng, C.C., SURFACE ULTRASONIC WAVE PARAMETRIC AMPLIFIER, IBM Tech Discl Bull v12 n10 p1699 (Mch 70)

Tseng, C.C., PIEZOELECTRIC SURFACE WAVES IN CUBIC CRYSTALS, J Appl Phys v41 n6 p2270 (May 70)

Tseng, C.C., INCREASED CAPACITY AND REDUCED CROSSTALK FOR CODED SURFACE-WAVE DELAY LINES USING ORTHOGONAL PAIRS OF COMPLEMENTARY SEQUENCE (Abstr.), IEEE Trans Sonics & US vSU-18 n1 p61 (Jan 71)

Tseng, C.C., SIGNAL MULTIPLEXING IN SURFACE-WAVE DELAY LINES USING ORTHOGONAL PAIRS OF GOLAY'S COMPLEMENTARY SEQUENCES, IEEE Trans Sonics & US vSU-18 n2 p103 (Apr 71)

Turner, C.W., Shomon, A., White, R.M., REDUCED-VOLTAGE OPERATION OF A SURFACE-ELASTIC-WAVE AMPLIFIER, Electron Lett v5 n11 p244 (29 May 69)

Tursonov, D.A., GENERALIZED SURFACE WAVES IN CUBIC CRYSTALS, Soviet Phys - Acoustics v13 n1 p78 (Jly/Sep 67)

Van den Heuvel, A.P., SURFACE-WAVE ELECTRONICS, Science and Technology n85 p52 (Jan 69)

Van den Heuvel, A.P., Owen D.B., Joshi, S.G., HYBRID TRANSDUCER: A NEW TECHNIQUE FOR GENERATING SURFACE WAVES ON NON-PIEZOELECTRIC SUBSTRATES (Abstr.), IEEE Trans Sonics & US vSU-17 n1 p56 (Jan 70)

Van de Vaart, H., Matthews, H., PROPAGATING MAGNETOELASTIC WAVES IN AN INFINITE PLATE, Appl Phys Lett v16 n4 p153 (15 Feb 70)

Van de Vaart, H., Matthews, H., MAGNETOELASTIC LOVE WAVES IN A MAGNETIC LAYERED NONMAGNETIC SUBSTRATE, Appl Phys Lett v16 n5 p222 (1 Mch 70)

Van de Vaart, H., DISPERSION CURVES FOR MAGNETOELASTIC LOVE WAVES IN METAL COATED LAYERED SUBSTRATES (Abstr.), IEEE Trans Sonics & US vSU-18 n1 p51 (Jan 71)

Van Duzer, T., LENSES AND GRADED FILMS FOR FOCUSING AND GUIDING ACOUSTIC SURFACE WAVES, IEEE Proc v58 n8 p1230 (Aug 70)

Vaskova V.I., Viktorov, I.A., INVESTIGATION OF THE AMPLIFICATION OF ULTRASONIC SURFACE WAVES IN A CADMIUM SULFIDE CRYSTAL, Soviet Phys - Acoustics v15 n4 p460 (Apr/Jne 70)

Verevkina, L.V., Merkulov, L.G., Tursanov, D.A., SURFACE WAVES IN A QUARTZ CRYSTAL, Soviet Phys - Acoustics v12 n3 p254 (Jan/Mch 67)

Viktorov, I.A., INVESTIGATION OF METHODS FOR EXCITING RAYLEIGH WAVES, Soviet Phys - Acoustics v7 n3 p236 (Jan/Mch 62)

Viktorov, I.A., RAYLEIGH WAVES IN THE ULTRASONIC RANGE, Soviet Phys - Acoustics v8 n3 p118 (Oct/Dec 62)

Viktorov,I.A.,"RAYLEIGH AND LAMB WAVES" Plenum Press, New York p5 (1967)

Viktorov, I.A., AMPLIFICATION OF RAYLEIGH WAVES IN GALLIUM ARSENIDE CRYSTALS, Soviet Phys - Acoustics v16 n1 p30 (Jly/Sep 70)

Vinogradov, K.N., Ul'yanov, G.K., MEASUREMENT OF THE VELOCITY AND ATTENUATION OF ULTRASONIC SURFACE WAVES IN HARD MATERIALS, Soviet Phys - Acoustics v5 n3 p296 (Feb 60)

Voltmer, F.W., Ippen, E.P., White, R.M., Lim, T.C., Farnell, G.W., MEASURED AND CALCULATED SURFACE-WAVES VELOCITIES, IEEE Proc v56 n9 p1634 (Sep 68)

Voltmer, F.W., White, R.M., Turner, C.W., MAGNETOSTRICTIVE GENERATION OF SURFACE ELASTIC WAVES, Appl Phys Lett v15 n5 p153 (1 Sep 69)

Vogel, R.F., Koerber, G.G., GENERAL SH-MODE PIEZOELECTRIC SURFACE WAVES (Abstr.), IEEE Trans Sonics & US vSU-18 n1 p48 (Jan 71)

Waldner, M., Pedinoff, M.E., Jones, W.R., Campbell, J.J., FUNDAMENTAL AND HIGHER-ORDER MODE SURFACE WAVE PROPAGATION ON LAYERED ANISOTROPIC SUBSTRATES - GOLD ON LITHIUM NIOBATE, J Appl Phys v41 n6 p2276 (May 70)

Waldron, R.A., SOME PROBLEMS IN THE THEORY OF GUIDED MICROSONIC WAVES, IEEE Trans Microwave T&T vMTT-17 n11 p893 (Nov 69)

Waldron, R.A., MODE SPECTRUM OF A MICROSOUND WAVEGUIDE CONSISTING OF AN ISOTROPIC RECTANGULAR OVERLAY ON A PERFECTLY RIGID SUBSTRATE, IEEE Trans Sonics & US vSU-18 n1 p8 (Jan 71)

Waldron, R.A., MICROSOUND WAVEGUIDES AND WAVEGUIDE COMPONENTS (Abstr.), IEEE Trans Sonics & US vSU-18 n1 p47 (Jan 71)

Wang, W.C., Staecker, P., Li, R.C.M., ELASTIC WAVES GUIDED BY A FLUID LAYER: A NEW TYPE OF ULTRASONIC WAVE PROPAGATION, Appl Phys Lett v16 n8 p291 (15 Apr 70)

Warner, A.W., Onoe, M., Coquin, G.A., DETERMINATION OF ELASTIC AND PIEZOELECTRIC CONSTANTS FOR CRYSTALS IN CLASS (3m), Ac Soc Am J v42 n6 p1223 (Jne 67)

Warner, A.W., Coquin, G.A., Meitzler, A.H., Fink, J.L., PIEZOELECTRIC PROPERTIES OF $Ba_2NaNb_5O_{15}$, Appl Phys Lett v14 n1 p34 (1 Jan 69)

Wauk, M.T., Winslow, D.K., DETERMINATION OF ACOUSTIC CONVERSION LOSS BY IMPEDANCE MEASUREMENTS, IEEE Trans Sonics & US vSU-16 n3 p86 (Jly 69)

Weglein, R.D., Pedinoff, M.E., Winston, H., DIFFRACTION SPREADING OF SURFACE WAVES ON $LiNbO_3$, Electron Lett v6 n20 p655 (1 Oct 70)

Weglein, R.D., Pedinoff, M.E., Winston, H., DIFFRACTION SPREADING OF SURFACE WAVES ON $LiNbO_3$, (Abstr.), IEEE Trans Sonics & US vSU-18 n1 p47 (Jan 71)

Welsh, F.S., SURFACE-WAVE TEMPERATURE COEFFICIENTS ON LITHIUM TANTALATE, IEEE Trans Sonics & US vSU-18 n2 p108 (Apr 71)

Wemple, S.H., DiDomenico, M., Jr., Camlibel, I., RELATIONSHIP BETWEEN LINEAR AND QUADRATIC ELECTRO-OPTICAL COEFFICIENTS IN $LiNbO_3$ AND OTHER OXYGEN-OCTAHEDRA FERROELECTRICS BASED ON DIRECT MEASUREMENT OF SPONTANEOUS POLARIZATION, Appl Phys Lett v12 n6 p209 (15 Mch 68)

Wen, C.F., Mayo, R.F., Schnitzler, P., MECHANICALLY VARIABLE ACOUSTIC SURFACE-WAVE DELAY LINE WITH A NONCONTACTING TRANSDUCER, Ac Soc Am J v41 n5 p1372 (May 67)

White, D.L., AMPLIFICATION OF ULTRASONIC WAVES IN PIEZOELECTRIC SEMICONDUCTORS, J Appl Phys v33 n8 p2547 (Aug 62)

White, D.L., SONIC CIRCUITS: A WAVE-GUIDE SYSTEM FOR ULTRASONIC SURFACE WAVES, (Abstr.), IEEE Trans Sonics & US vSU-15 n1 p76 (Jan 68)

White, R.M., Voltmer, F.W., DIRECT PIEZOELECTRIC COUPLING TO SURFACE ELASTIC WAVES, Appl Phys Lett v7 n12 p314 (15 Dec 65)

White, R.M., Voltmer, F.W., ULTRASONIC SURFACE-WAVE AMPLIFICATION IN CADMIUM SULFIDE, Appl Phys Lett v8 n2 p40 (15 Jan 66)

White, R.M., SURFACE ELASTIC-WAVE PROPAGATION AND AMPLIFICATION, IEEE Trans Electron Dev vED-14 n4 p181 (Apr 67)

White, R.M., SURFACE ELASTIC WAVES, IEEE Proc v58 n8 p1238 (Aug 70)

Whitehouse, H.J., A SURFACE-WAVE MATCHED FILTER FOR SPECTRUM ANALYSIS, (Abstr.) IEEE Trans Sonics & US vSU-15 n1 p59 (Jan 68)

Whitehouse, H.J., A WIDE-BAND LOW-INSERTION LOSS MULTI-ELECTRODE SURFACE-WAVE TRANSDUCER (Abstract), IEEE Trans Sonics & US vSU-17 n1 p69 (Jan 70)

Whiteman, R.L., Laub, L.J., Bates, W.J., ACOUSTIC SURFACE DISPLACEMENT MEASUREMENT ON A WEDGE-SHAPED TRANSDUCER USING OPTICAL PROBE TECHNIQUE, IEEE Trans Sonics & US vSU-15 n3 p186 (Jly 68)

Willis, W., CONSIDERATION OF SMALL-SCALE EFFECTS IN PREDICTING THE RESPONSES OF INTERDIGITAL ACOUSTIC SURFACE-WAVE TRANSDUCERS (Abstr.), IEEE Trans Sonics & US vSU-18 n1 p60 (Jan 71)

Worley, J.C., Matthews, H., BROADBAND UNIDIRECTIONAL SURFACE-WAVE TRANSDUCER, (Abstr.), IEEE Trans Sonics & US vSU-18 n1 p52 (Jan 71)

Yip, G.L., LAUNCHING EFFICIENCY OF THE HE II SURFACE WAVE MODE ON A DIELECTRIC ROD, IEEE Trans Microwave T&T vMTT-18 n12 p1033 (Dec 70)

Yoder, M.M., CHALLENGE OF SEVENTIES: SYSTEMS EXPLOITATION OF THE MULTIFUNCTIONAL PROPERTIES OF PRAETERSONICS TECHNOLOGY, Proc 20th Electronic Components Conf p455 (May 13-15 70)

Yoshida, K., Yamanishi, M., INTERACTION BETWEEN SURFACE ELASTIC WAVES AND DRIFTING CARRIERS IN LAYERED SYSTEMS, Japan J Appl Phys v7 n9 p1143 (Sep 68)

Young, P., EFFECT OF BOUNDARY CONDITIONS ON THE PROPAGATION OF SURFACE MAGNETO-STATIC WAVES IN A TRANSVERSELY MAGNETIZED THIN Y.I.G. SLAB, Electron Lett v5 n18 p429 (4 Sep 69)

Zemon, S., Zucker, J., Wasko, J.H., Conwell, E.M., Ganguly, A.K., PARAMETRIC AMPLIFICATION OF ULTRASONIC WAVES IN CdS, Appl Phys Lett v12 n11 p378 (1 Jne 68)

Zemon, S., Conwell, E.M., EFFECTS OF SURFACE STATES ON SURFACE WAVE AMPLIFICATION, Appl Phys Lett v17 n5 p218 (1 Sep 70)

ACOUSTO-OPTICS

Abrams, R.L., Pinnow, D.A., ACOUSTO-OPTIC PROERTIES OF CRYSTALLINE GERMANIUM, J Appl Phys v41 n7 p2765 (Jne 70)

Abrams, R.L., Pinnow, D.A., EFFICIENT ACOUSTOOPTIC MODULATION AT 3.39 AND 10.6 μm IN CRYSTALLINE GERMANIUM, IEEE J Quantum Electronics vQE-7 n3 p135 (Mch 71)

Addison, R.C., Auld, B.A., Wilkinson, J.H., ELECTRICALLY CONTROLLED ACOUSTIC BEAM DEFLECTION, IEEE Proc v55 p68 (Jan 67)

Adler, R., INTERACTION BETWEEN LIGHT AND SOUND, IEEE Spectrum v4 p42 (May 67)

Adler, R., INTERACTION BETWEEN LIGHT AND SOUND: AN INTRODUCTION, 1969 IEEE Int'l Convention Digest p64 (Mch 69)

Adler, R., ACOUSTOOPTICS: EFFECTS AND APPLICATIONS (Abstr.), IEEE Trans Sonics & US vSU-18 n1 p58 (Jan 71)

Atzeni, C., Pantani, L., OPTICAL SIGNAL-PROCESSING THROUGH DUAL-CHANNEL ULTRA-SONIC LIGHT MODULATORS, IEEE Proc v58 n3 p501 (Mch 70)

Auld, B.A., Wilson, D.A., BRAGG SCATTERING OF INFRARED RADIATION FROM COHERENT SPIN WAVES, J Appl Phys v38 p3331 (Jly 67)

Auld, B.A., Webb, D.C., Winslow, D.K., ACOUSTIC BEAM SCANNING BY OPTICAL CONTROL OF PIEZOELECTRIC COUPLING, IEEE Proc v57 p713 (Apr 69)

Bennett, B.I., Maradudin, A.A., Swanson, L.R., THEORY OF THE BRILLOUIN SCATTERING OF LIGHT BY ACOUSTIC PHONONS IN A METAL (Abstr.), IEEE Trans Sonics & US vSU-18 n1 p57 (Jan 71)

Berger, H., ULTRASONIC IMAGING SYSTEMS FOR NONDESTRUCTIVE TESTING, Ac Soc Am J v45 n4 p859 (Apr 69)

Berry, M.V., "THE DIFFRACTION OF LIGHT BY ULTRASOUND", New York: Academic Press (1966)

Boersma, S.L., HOLOGRAPHIC TELEVISION, IEEE Proc v58 n5 p803 (May 70)

Borsuk, G.M., Thaler, W.J., FREQUENCY-MODULATED LASER COMMUNICATION SYSTEM, IEEE Trans Sonics & US vSU-17 n4 p207 (Oct 70)

Brienza, M.J., DeMaria, A.J., CONTINUOUSLY-VARIABLE ULTRASONIC-OPTICAL DELAY LINE, Appl Phys Lett v9 p312 (Oct 66)

Brienza, M.J., VARIABLE TIME COMPRESSION, EXPANSION, AND REVERSAL OF R.F. SIGNALS BY LASER-ACOUSTIC TECHNIQUES, Appl Phys Lett v12 p181 (Mch 68)

Brown, B.R., Rowe, S.H., ACOUSTIC LIGHT DEFLECTOR CHROMATIC VARIATION COMPENSATION, IBM Tech Discl Bull v13 n9 p2634 (Feb 71)

Bryngdahl, O., FREQUENCY MODULATION OF A LIGHTBEAM BY USE OF ACOUSTICAL FIELDS, IBM Tech Discl Bull v13 n2 p514 (Jly 70)

Caddes, D.E., Osterink, L.M., Targ, R., MODE LOCKING OF THE CO_2 LASER, Appl Phys Lett v12 p74 (Feb 68)

Chang, I.C., Lean, E.G., SPATIALLY MODULATED LASER, IBM Tech Discl Bull v13 n2 p419 (Jly 70)

Chang, W.S.C., ACOUSTOOPTICAL DEFLECTION IN THIN FILMS, IEEE J Quantum Electronics vQE-7 n4 p167 (Apr 71)

Chiao, R.Y., Townes, C.H., Stoicheff, B.P., STIMULATED BRILLOUIN SCATTERING AND COHERENT GENERATION OF INTENSE HYPERSONIC WAVES, Phys Rev Lett v12 p592 (May 64)

Chu, R.S., Tamir, T., GUIDED-WAVE THEORY OF LIGHT DIFFRACTION BY ACOUSTIC MICROWAVES, IEEE Trans Microwave T&T vMTT-17 p1002 (Nov 69)

Chu, R.S., Tamir, T., GUIDED-WAVE THEORY OF LIGHT DIFFRACTION BY ACOUSTIC MICROWAVES (Corrected reprint), IEEE Trans Microwave T&T vMTT-18 n8 p486 (Aug 70)

Cohen, M.G., Gordon, E.I., ELECTRO-OPTIC (KTN) GRATINGS FOR LIGHT BEAM MODULATION AND DEFLECTION, Appl Phys Lett v5 n9 p181 (1 Nov 64)

Cohen, M.G., Gordon, E.I., ACOUSTIC BEAM PROBING USING OPTICAL TECHNIQUES, Bell Syst Tech J v44 p693 (Apr 65)

Cohen, M.G., OPTICAL STUDY OF ULTRASONIC DIFFRACTION AND FOCUSING IN ANISOTROPIC MEDIA, J Appl Phys v38 p3821 (Sep 67)

Collins, J.H., Lean, E.G.H., Shaw, H.J., PULSE COMPRESSION BY BRAGG DIFFRACTION OF LIGHT WITH MICROWAVE SOUND, Appl Phys Lett v11 p240 (Oct 67)

Collins, J.H., Wilson, D.A., OPTICAL PROBING OF MAGNETOSTATIC MODES IN Y.I.G. DELAY LINES, Appl Phys Lett v12 p331 (May 68)

Coquin, G.A., Griffin, J.P., Anderson, L.K., WIDE-BAND ACOUSTO-OPTIC DEFLECTORS USING ACOUSTIC BEAM STEERING, IEEE Trans Sonics & US vSU-17 n1 p34 (Jan 70)

Coquin, G.A., ACOUSTOOPTIC INTERACTIONS IN PIEZOELECTRIC CRYSTALS (Abstr.), IEEE Trans Sonics & US vSU-17 nl p66 (Jan 70)

Coquin, G.A., Pinnow, D.A., Warner, A.W., PHYSICAL PROPERTIES OF LEAD MOLYBDATE RELEVANT TO ACOUSTOOPTIC DEVICE APPLICATIONS (Abstr.), IEEE Trans Sonics & US vSU-18 nl p57 (Jan 71)

Coquin, G.A., Pinnow, D.A., Warner, A.W., PHYSICAL PROPERTIES OF LEAD MOLYBDATE RELEVANT TO ACOUSTO-OPTIC DEVICE APPLICATIONS, J Appl Phys v42 n6 p2162 (May 71)

Cummins, H.Z., Knable, N., SINGLE-SIDEBAND MODULATION OF COHERENT LIGHT BY BRAGG REFLECTION FROM ACOUSTICAL WAVES, IEEE Proc v51 p1246 (Sep 63)

De Maria, A.J., Gagosz, R., ULTRASONIC FEEDBACK MODULATION OF AN OPTICAL MASER OSCILLATOR, IEEE Proc v50 p1522 (Jne 62)

De Maria, A.J. Gagosz, R., Barnard, G., ULTRASONIC REFRACTION SHUTTER FOR OPTICAL MASER OSCILLATORS, J Appl Phys v34 p453 (Mch 63)

De Maria, A.J., ULTRASONIC DIFFRACTION SHUTTER FOR OPTICAL MASER OSCILLATOR, J Appl Phys v34 p2984 (Oct 63)

Dixon, R.W., Cohen, M.G., A NEW TECHNIQUE FOR MEASURING MAGNITUDES OF PHOTO-ELASTIC TENSORS AND ITS APPLICATION TO LITHIUM NIOBATE, Appl Phys Lett v8 p205 (Apr 66)

Dixon, R.W., GORDON, E.I., ACOUSTIC LIGHT MODULATORS USING OPTICAL HETERODYNE MIXING, Bell Syst Tech J v46 p367 (Feb 67)

Dixon, R.W., Matthews, H., DIFFRACTION OF LIGHT BY ELASTIC WAVES IN Y.I.G., Appl Phys Lett v10 p195 (Apr 67)

Dixon, R.W., ACOUSTIC DIFFRACTION OF LIGHT IN ANISOTROPIC MEDIA, IEEE J Quantum Electronics vQE-3 p85 (Feb 67)

Dixon, R.W., ACOUSTIC NONLINEAR FREQUENCY MIXING DETECTED USING OPTICAL BRAGG DIFFRACTION, Appl Phys Lett v11 1340 (Dec 67)

Dixon, R.W., OPTICAL INVESTIGATION OF MAGNETICALLY INDUCED ELASTIC-WAVE DISPERSION IN Y.I.G., J Appl Phys v38 p3634 (Aug 67)

Dixon, R.W., PHOTOELASTIC PROPERTIES OF SELECTED MATERIALS AND THEIR RELEVANCE FOR APPLICATIONS TO ACOUSTIC LIGHT MODULATORS AND SCANNERS, J Appl Phys v38 p5149 (Dec 67)

Dixon, R.W., ACOUSTOOPTIC INTERACTIONS AND DEVICES, IEEE Trans Electron Dev vED-17 n3 p229 (Mch 70)

Farah, O.G., Wagner, P.E., Cummins, H.Z., BRILLOUIN SCATTERING FROM ULTRASONIC MICROWAVE PULSES IN $LiNbO_3$, A Appl Phys v40 p2365 (May 69)

Flinchbaugh, D.E., Crofut, W.A., AN L-BAND ACOUSTIC DIFFRACTION GRATING DELAY DEVICE FOR BROADBAND PULSE COMPRESSION, IEEE Proc v55 p2183 (Dec 67)

Flinchbaugh, D.E., ACOUSTO-OPTIC TECHNIQUES IN LASER COMMUNICATIONS, Opt Spectra v4 n8 p49 (Sep 70)

Foster, L.C., Ewy, M.D., Crundy, C.B., LASER MODE LOCKING BY AN EXTERNAL DOPPLER CELL, Appl Phys Lett v6 p6 (Jan 65)

Foster, T.M., DYNAMIC RANGE OF OPTOACOUSTIC CORRELATIONS (Abstr.), IEEE Trans Sonics & US vSU-18 n1 p56 (Jan 71)

Gericke, O.R., Grubinskas, R.C., UTILIZATION OF THE LIQUID SURFACE LEVITATION EFFECT AS A MEANS OF ULTRASONIC IMAGE CONVERSION FOR MATERIALS INSPECTION, Ac Soc Am J v45 n4 p872 (Apr 69)

Girrbach, H.F., DUAL WATER CELL, IBM Tech Discl Bull v13 n10 p2851 (Mch 71)

Gordon, E.I., A REVIEW OF ACOUSTOOPTICAL DEFLECTION AND MODULATION DEVICES, Appl Opt v5 p1629 (Oct 66), also IEEE Proc v54 p1391 (Oct 66)

Gordon, E.I., FIGURE OF MERIT FOR ACOUSTO-OPTICAL DEFLECTION AND MODULATION DEVICES, IEEE J Quantum Electronics vQE-2 p104 (May 66)

Gordon, R.L., STIMULATED BRILLOUIN SCATTERING IN PIEZOELECTRIC SEMICONDUCTORS, J Appl Phys v39 p306 (Jan 68)

Gottlieb. M., Conroy, J., Foster, T., LARGE TIME-BANDWIDTH OPTOACOUSTIC SIGNAL PROCESSING DEVICES (Abstr.), IEEE Trans Sonics & US vSU-18 n1 p57 (Jan 71)

Hance, H.V., Parks, J.K., WIDE-BAND MODULATION OF A LASER BEAM, USING BRAGG ANGLE DIFFRACTION BY AMPLITUDE-MODULATED ULTRASONIC WAVES, Ac Soc Am j v38 p14 (Jly 65)

Hargrove, I.E., Fork, R.L., Pollack, M.A., LOCKING OF He-Ne MODES INDUCED BY SYNCHRONOUS INTRACAVITY MODULATION, Appl Phys Lett v5 p4 (Jly 64)

Harris, E.P., Lean, E.G.H., Powell, C.B., INTERNAL MODULATION OF INJECTION LASERS USING ACOUSTIC WAVES, IBM Tech Discl Bull v13 n12 p3871 (May 71)

Harris, S.E., Nieh, S.T.K., Winslow, D.K., ELECTRONICALLY TUNABLE ACOUSTO-OPTIC FILTER, Appl Phys Lett v15 p325 (Nov 69)

Harris S.E., NEW TYPE OF ELECTRONICALLY TUNABLE OPTICAL FILTER (Abstr.), IEEE Trans Sonics & US vSU-18 n1 p56 (Jan 71)

Havlice, J.F., VISUALIZATION OF ACOUSTIC BEAMS USING LIQUID CRYSTALS, Electron Lett v5 p477 (Oct 69)

Havlice, J.F., Bond, W.L., Wigton, L.B., "ELASTIC" POYNTING VECTOR IN A PIEZO-ELECTRIC MEDIUM, IEEE Trans Sonics & US vSU- 17 n4 p246 (Oct 70)

Heiniche, W., Winterling, G., THE TEMPERATURE DEPENDENCE OF STIMULATED BRILLOUIN SCATTERING IN QUARTZ, Appl Phys Lett v11 p231 (Oct 67)

Henderson, D.M., Abrams, R.L., A COMPARISON OF ACOUSTOOPTIC AND ELECTROOPTIC MODULATORS AT 10.6 MICRONS, Opt Commun v2 n5 p223 (Oct 70)

Ishida, A., Sumi, M., ACOUSTO-OPTICAL MEASUREMENT OF ACOUSTIC-WAVE GENERATION IN PIEZOELECTRICS COUPLED TO A GUNN OSCILLATOR, Appl Phys Lett v18 n6 p252 (15 Mch 71)

Johnson, V.R., Olson, F.A., PHOTOELASTIC PROPERTIES OF Y.A.G., IEEE Proc v55 p709 (May 67)

Key, P.Y., Harrison, R.G., Little, V.I., Katzenstein, J., BRAGG REFLECTION FROM A PHASE GRATING INDUCED BY NONLINEAR OPTICAL EFFECTS IN LIQUIDS, IEEE J Quantum Electronics vQE-6 n10 p641 (Oct 70)

Klein, W.R., Cook, B.D., UNIFIED APPROACH TO ULTRASONIC LIGHT DIFFRACTION, IEEE Trans Sonics & US vSU-14 p123 (Jly 67)

Korpel, A., Adler, R., Desmares, P., Smith, T.M., AN ULTRASONIC LIGHT DEFLECTION SYSTEM, IEEE J Quantum Electronics vQE-1 p60 (Apr 65)

Korpel, A., Adler, R., Desmares, P., Watson, W., A TELEVISION DOSPLAY USING ACOUSTIC DEFLECTION AND MODULATION OF COHERENT LIGHT, IEEE Proc v54 p1429 (Oct 66)

Korpel, A., ACOUSTIC IMAGING BY DIFFRACTED LIGHT: I - TWO DIMENSIONAL INTERACTION IEEE Trans Sonics & US vSU-15 p153 (Jly 68)

Korpel, A., VISUALIZATION OF THE CROSS-SECTION OF A SOUND BEAM BY BRAGG DIFFRACTION OF LIGHT, Appl Phys Lett v9 p425 (Dec 66)

Korpel, A., Deasmares, P., RAPID SAMPLING OF ACOUSTICAL HOLOGRAMS BY LASER-SCANNING TECHNIQUES, Ac Soc Am J v45 n4 p881 (Apr 69)

Korpel, A., ASTIGMATIC IMAGING PROPERTIES OF BRAGG DIFFRACTION, Ac Soc Am J v49 n3(Pt.3) p1059 (Mch 71)

Krause, J.T., Kurkjian, C.R., Pinnow, D.A., Sigety, E.A., LOW ACOUSTIC LOSS CHALCOGENIDE GLASSES - A NEW CATEGORY OF MATERIALS FOR ACOUSTIC AND ACOUSTO-OPTIC APPLICATIONS, Appl Phys Lett v17 n9 p367 (1 Nov 70)

Krause, J.T., Kurkjian, C.R., Sigety, E.A., LOW ACOUSTIC LOSS CHALCOGENIDE GLASSES FOR ACOUSTO-OPTIC AND ULTRASONIC DEVICES (Abstr.), IEEE Trans Sonics & US vSU-18 n1 p49 (Jan 71)

Krischer, C., ULTRASONIC ATTENUATION MEASUREMENTS IN FUSED SILICA USING BRAGG DIFFRACTION (Abstr.), IEEE Trans Sonics & US vSU-17 n1 p67 (Jan 70)

La Macchia, J.T., Coquin, G.A., SIMULTANEOUS X, Y, ACOUSTOOPTIC DEFLECTION, IEEE Proc v59 n2 p304 (Feb 71)

Landry, J., Powers, J., Wade, G., ULTRASONIC IMAGING OF INTERNAL STRUCTURE BY BRAGG DIFFRACTION, Appl Phys Lett v15 p186 (Sep 69)

Lean, E.G.H., Quate, C.F., Shaw, H.J., CONTINUOUS DEFLECTION OF LASER BEAMS, Appl Phys Lett v10 p48 (Jan 67)

Lee, S.H., Milnes, A.G., A PROPOSED METHOD OF IMPROVING THE RATE OF ACOUSTOOPTICAL SCANNING OF A LASER BEAM, IEEE Proc v57 p2158 (Dec 69)

Maloney, W.T., Carleton, H.R., LIGHT DIFFRACTION BY TRANSVERSE ULTRASONIC WAVES IN HEXAGONAL CRYSTALS, IEEE Trans Sonics & US vSU-14 p135 (Jly 67)

Maloney, W.T., Meltz, G., Gravel, R., OPTICAL PROBING OF THE FRESNEL AND FRAUNHOFER REGIONS OF A RECTANGULAR ACOUSTIC TRANSDUCER, IEEE Trans Sonics & US vSU-15 p167 (Jly 68)

Maloney, W.T., ACOUSTO-OPTICAL APPROACHES TO RADAR INFORMATION PROCESSING, 1969 IEEE Int'l Conv Digest p72 (Mch 69)

Maloney, W.T., Gravel, R.L., MEASUREMENT OF 'SHEAR' PHOTOELASTIC CONSTANTS IN LITHIUM NIOBATE, IEEE Proc v57 p1232 (Jly 69)

Maloney, W.T., ACOUSTOOPTICAL APPROACHES TO RADAR SIGNAL PROCESSING, IEEE Spectrum v6 n10 p40 (Oct 69)

Maydan, D., ACOUSTOOPTICAL PULSE MODULATORS, IEEE J Quantum Electronics vQE-6 n1 p15 (Jan 70)

Maydan, D., Sealer, D.A., ACOUSTOOPTIC MODULATORS WITH NANOSECOND RISE TIME CAPABILITY (Abstr.), IEEE Trans Sonics & US vSU-17 n1 p65 (Jan 70)

Mazumder, M.K., LASER DOPPLER VELOCITY MEASUREMENT WITHOUT DIRECTIONAL AMBIGUITY BY USING FREQUENCY SHIFTED INCIDENT BEAMS, Appl Phys Lett v16 n11 p462 (1 Jne 70)

McMahon, D.H., A COMPARISON OF BRILLOUIN SCATTERING TECHNIQUES FOR MEASURING MICROWAVE ACOUSTIC ATTENUATION, IEEE Trans Sonics & US vSU-14 p103 (Jly 67)

McMahon, D.H., RELATIVE EFFICIENCY OF OPTICAL BRAGG DIFFRACTION AS A FUNCTION OF INTERACTION GEOMETRY, IEEE Trans Sonics & US vSU-16 p41 (Apr 69)

Papoular, M., Generation of tunable, high frequency hypersounds, Solid State Commun v4 p129 (Mch 66)

Parks, J.K., Tsai, C.S., THERMAL EFFECTS OBSERVED IN ULTRASONIC DIFFRACTION EXPERIMENTS WITH RUTILE CRYSTALS, IEEE Proc v55 p701 (May 67)

Pinnow, D.A., Dixon, R.W., ALPHA-IODIC ACID: A SOLUTION-GROWN CRYSTAL WITH A HIGH FIGURE OF MERIT FOR ACOUSTO-OPTIC DEVICE APPLICATIONS, Appl Phys Lett v13 p156 (Sep 68)

Pinnow, D.A., Van Uitert, L.G., Warner, A.W., Bonner, W.A., LEAD MOLYBDATE: A MELT-GROWN CRYSTAL WITH A HIGH FIGURE OF MERIT FOR ACOUSTO-OPTIC DEVICE APPLICATIONS, Appl Phys Lett, v15 p83 (Aug 69)

Pinnow, D.A., ACOUSTOOPTIC MATERIALS AND LIGHT DEFLECTION DEVICES (Abstr.), IEEE Trans Sonics & US vSU-17 n1 p65 (Jan 70)

Pinnow, D.A., GUIDE LINES FOR THE SELECTION OF ACOUSTOOPTIC MATERIALS, IEEE j Quantum Electronics vQE-6 n4 p223 (Apr 70)

Pinnow, D.A., Abrams, R.L., CHARACTERIZATION AND APPLICATION OF CRYSTALLINE GERMANIUM AS AN INFRARED ACOUSTOOPTIC MEDIUM (Abstr.), IEEE Trans Sonics & US vSU-18 n1 p57 (Jan 71)

Quate, C.F., Wilkinson, C.D., Winslow, D.K., INTERACTION OF LIGHT AND MICROWAVE SOUND, IEEE Proc v53 p1604 (Oct 65)

Powers, J.P., Wade, G., Wang, K., ACOUSTIC VELOCITY MEASUREMENT BY BRAGG IMAGING (Abstr.), IEEE Trans Sonics & US vSU-18 n1 p57 (Jan 71)

Randolph, J., Morrison, J., MODULATION TRANSFER CHARACTERISTICS OF AN ACOUSTO-OPTIC DEFLECTOR, Appl Opt v10 n6 p1383 (Jne 71)

Reintjes, J., Schulz, M.B., PHOTOELASTIC CONSTANTS OF SELECTED ULTRASONIC DELAY-LINE CRYSTALS, J Appl Phys v39 p5254 (Oct 68)

Rheault, F., MICROWAVE ACOUSTIC PROPAGATION OF FLUIDS BY BRAGG SCATTERING, Ac Soc Am J v49 n5(Pt.2) p1448 (May 71)

Ripper, J.E., ANALYSIS OF FREQUENCY MODULATION OF JUNCTION LASERS BY ULTRASONIC WAVES, IEEE J Quantum Electronics vQE-6 n2 p129 (Feb 70)

Rowe, S.H., FAST FLYBACK HIGH-SPEED OPTICAL SCANNER, IBM Tech Discl Bull v13 n2 p391 (Jly 70)

Sadagopan, V., Kuhn, L., ACOUSTO-OPTIC DEFLECTOR, IBM Tech Discl Bull v13 n1 p98 (Jne 70)

Schmitt, H.J., DIFFRACTION OF ELECTROMAGNETIC WAVES BY SOUND WAVES, Ac Soc Am J v33 b10 p1288 (Oct 61)

Sette, D., ULTRASONIC LENSES OF PLASTIC MATERIALS, Ac Soc Am J v21 n4 p375 (Jly 69)

Shen, Y.R., Bloembergen, N., INTERACTION BETWEEN LIGHT WAVES AND SPIN WAVES, Phys Rev v143 p372 (Mch 66)

Shulz, M.B., Holland, M.G., Davis, L., Jr., OPTICAL PULSE COMPRESSION USING BRAGG SCATTERING BY ULTRASONIC WAVES, Appl Phys Lett v11 p237 (Oct 67)

Siegman, A.E., Quate, C.F., Bjorkholm, J., Francois, G., FREQUENCY TRANSLATION OF AN He-Ne LASER'S OUTPUT FREQUENCY BY ACOUSTIC OUTPUT COUPLING INSIDE THE RESONANT CAVITY, Appl Phys Lett v5 p1 (Jly 64)

Singer, B., Tamir-Berman, T., DIFFRACTION OF LIGHT FROM A FILAMENTARY SOURCE BY PARALLEL ACOUSTIC WAVE FRONTS, Opt Soc Am J v60 n12 p1641 (Dec 70)

Smith, A.W., DIFFRACTION OF LIGHT BY MAGNETOELASTIC WAVES, Appl Phys Lett v11 p7 (Jly 67)

Smith, A.W., DIFFRACTION OF LIGHT BY MAGNETOELASTIC WAVES, IEEE Trans Sonics & US vSU-15 p161 (Jly 68)

Smith, A.W., OPTICAL PROBING OF MAGNETOELASTIC WAVES, Phys Rev Lett v20 p334 (Feb 68)

Smith, R.A., Wada, G., Powers, J., Landry, J., STUDIES OF RESOLUTION IN A BRAGG IMAGING SYSTEM, Ac Soc Am J v49 n3(Pt.3) p1062 (Mch 71)

Smith, T.M., Korpel, A., MEASUREMENT OF LIGHT-SOUND INTERACTION EFFICIENCIES IN SOLIDS, IEEE J Quantum Electronics vQE-1 p283 (Sep 65)

Spencer, E.G., Lenzo, P.V., Nassau, K., OPTICAL INTERACTION WITH ELASTIC WAVES IN LITHIUM NIOBATE, IEEE J Quantum Electronics vQE-2 p69 (Mch 66)

Spencer, E.G., Lenzo, P.V., Ballman, A.A., DIELECTRIC MATERIALS FOR ELECTROOPTIC, ELASTOOPTIC, AND ULTRASONIC DEVICE APPLICATIONS, IEEE Proc v55 p2074 (Dec 67)

Spencer, E.G., Guggenheim, H.J. Kominiak, G.J., BARIUM MANGANESE FLUORIDE, A NEW CRYSTAL FOR MICROWAVE ULTRASONICS, Appl Phys Lett v17 n7 p300 (1 Oct 70)

Squire, W.D., Whitehouse, H.J., Alsup, J.M., LINEAR SIGNAL PROCESSING AND ULTRASONIC TRANSVERSAL FILTERS, IEEE Trans Microwave T&T vMTT-17 p1020 (Nov 69)

Stark, H., King, M., Arm, M., LINEAR SPATIAL FILTERING WITH CROSSED ULTRASONIC LIGHT MODULATORS, IEEE Proc p1456 (Aug 69)

Steenhuysen, L.W.G., Termeulen, J., HOLOGRAPHIC RECORDING OF ACOUSTIC FIELDS IN TRANSPARENCIES BY MEANS OF A FREQUENCY MODULATED REFERENCE BEAM, Appl Opt v10 n4 p981 (Apr 71)

Streifer, W., Whinnery, J.R., ANALYSIS OF A DYE LASER TUNED BY ACOUSTO-OPTIC FILTER, Appl Phys Lett v17 n8 p335 (15 Oct 70)

Sweeney, H.E., Budd, W.E., THE NUMBER OF RESOLVABLE SPOTS IN A PHOTOELASTIC BEAM DEFLECTION SYSTEM, IEEE Proc v58 n7 p1162 (Jly 70)

Thurstone, F.L., HOLOGRAPHIC IMAGING WITH ULTRASOUND, Ac Soc Am J v45 n4 p895 (Apr 69)

Torguet, R., Dieulesaint, E., CONTINUOUS DEFLECTION OF LIGHT WITH LONGITUDINAL ACOUSTIC WAVES (Abstr.), IEEE Trans Sonics & US vSU-17 n1 p66 (Jan 70)

Tsai, C.S., Hance, H.V., OPTICAL IMAGING OF THE CROSS SECTION OF A MICROWAVE ACOUSTIC BEAM IN RUTILE BY BRAGG DIFFRACTION OF A LASER BEAM, Ac Soc Am J v42 n6 p1345 (Dec 67)

Tsai, C.S., Hance, H.V., EXPERIMENTAL INVESTIGATION OF THE RESOLUTION CAPABILITY OF MICROWAVE ULTRASONIC BEAM VISUALIZATION TECHNIQUES USING BRAGG DIFFRACTION OF A LASER BEAM, Ac Soc Am J v48 n5(Pt.2) p1110 (Nov 70)

Tsai, C.S., THE INCREASE OF BRAGG DIFFRACTION INTENSITY DUE TO ACOUSTIC RESONANCE AND ITS APPLICATION FOR DEMULTIPLEXING AND MULTIPLEXING IN LASER COMMUNICATION, Appl Opt v10 n1 p215 (Jan 71)

Uchida, N., Ohmachi, Y., ELASTIC AND PHOTOELASTIC PROPERTIES OF TeO_2 SINGLE CRYSTAL, J Appl Phys v40 p4692 (Nov 69)

Wade, G., Powers, J.P., DeSouza, A.A., EXPERIMENTAL DETECTION OF EVANESCENT ULTRASONIC WAVES BY BRAGG DIFFRACTION OF LIGHT, Ac Soc Am J v45 n5 p1247 (May 69)

Wade, G., Landry, J., Powers, J., ADVANCES IN ULTRASONIC IMAGING BY BRAGG DIFFRACTION (Abstr.), IEEE Trans Sonics & US vSU-17 n1 p57 (Jan 70)

Wang, S., Thomas, G., SPIN-WAVE INTERACTIONS WITH A LASER BEAM IN Y.I.G., Appl Phys Lett v10 p210 (Apr 67)

Warner, A.W., Sittig, E.K., RESOLUTION LIMITATIONS ON ACOUSTO-OPTIC LIGHT DEFLECTORS IMPOSED BY THERMAL EFFECTS (Abstr.), IEEE Trans Sonics & US vSU-18 n1 p57 (Jan 71)

Watson, W.H., Korpel, A., EQUALIZATION OF ACOUSTOOPTIC DEFLECTION CELLS IN A LASER COLOR TV SYSTEM, Appl Opt v9 n5 p1176 (May 70)

Willard, G.W., ULTRA-SOUND WAVES MADE VISIBLE, Bell Lab Rec v25 n5 p194 (May 47)

Willard, G.W., CRITERIA FOR NORMAL AND ABNORMAL ULTRASONIC LIGHT DIFFRACTION EFFECTS, Ac Soc Am J v21 n2 p101 (Mch 49)

Willard, G.W., FOCUSING ULTRASONIC RADIATORS, Ac Soc Am J v21 n4 p360 (Jly 49)

Wu, W.H., CALCULATION OF DEFLECTION EFFICIENCY IN A PHASE-MODULATED ACOUSTO-OPTICAL DEFLECTOR, Appl Opt v9 n2 p506 (Feb 70)

Yariv, A., QUANTUM THEORY FOR PARAMETRIC INTERACTIONS OF LIGHT AND HYPERSOUND, IEEE J Quantum Electronics vQE-1 n1 p28 (Apr 65)

****

PATENTS

PATENTS RELATED TO ACOUSTIC SURFACE WAVE DEVICES, PUBLISHED BETWEEN 1962 - 1971.

3,070,761 (Dec. 25, 1962)
A.B.C. Rankin, assignor to S. Smith & Sons (England) Ltd., London, England,
ULTRASONIC DELAY LINES.

3,289,114 (Nov. 29, 1966)
J.H. Rowen, assignor to Bell Telephone Laboratories, Inc., New York,
TAPPED ULTRASONIC DELAY LINE AND USES THEREFOR.

3,353,120 (Nov. 14, 1967)
P. Tournois, assignor to CSF - Compagnie Generale de Telegraphie Sans Fil, France,
ACOUSTIC PROPAGATION LINE FOR COMPRESSING TRAINS OF ELECTRIC WAVES.

3,360,749 (Dec. 26, 1967)
E.K. Sittig, assignor to Bell Telephone Laboratories, Inc., New York,
ELASTIC WAVE DELAY DEVICE.

3,376,572 (Apr. 2, 1968)
R.F. Mayo, assignor to Radio Corporation of America.
ELECTROACOUSTIC WAVE SHAPING DEVICE.

3,378,793 (Apr. 16, 1968)
W.S. Mortley, assignor to The Marconi Company, Ltd., London, England,
DISPERSIVE DELAY CELLS.

3,388,334 (June 11, 1968)
R. Adler, assignor to Zenith Radio Corporation, Chicago, Illinois,
SOLID STATE TRAVELING WAVE DEVICES.

3,400,341 (Sept. 3, 1968)
E.K. Sittig, assignor to Bell Telephone Laboratories, Inc., New York,
ELASTIC DISPERSIVE DELAY LINE.

3,406,358 (Oct. 15, 1968)
H. Seidel, and D.L. White, assignors to Bell Telephone Laboratories, New York,
ULTRASONIC SURFACE WAVEGUIDES.

3,409,848 (Nov. 5, 1968)
A.H. Meitzler, and H.F. Tiersten, assignors to Bell Telephone Laboratories, N.Y.,
ELASTIC SURFACE WAVEGUIDE.

1,151,735 (May 14, 1969) British Patent,
H. Jouffroy, and J. Desbois, assignors to Thomson CSF, Paris, France,
DEVICE FOR EXCITING OR RECOVERING SURFACE WAVES.

3,446,974 (May 27, 1969
R.F. Seiwatz, assignor to Zenith Radio Corporation, Chicago,
SOLID STATE ACOUSTIC SIGNAL TRANSLATING DEVICE WITH LIGHT ACTIVATED ELECTRODE INTERCONNECTIONS.

3,446,975 (May 27, 1969)
R. Adler and A.J. DeVries, assignors to Zenith Radio Corporation, Chicago,
ACOUSTO-ELECTRIC FILTER UTILIZING SURFACE WAVE PROPAGATION IN WHICH THE CENTER FREQUENCY IS DETERMINED BY A CONDUCTIVITY PATTERN RESULTING FROM AN OPTICAL IMAGE.

3,464,033 (Aug. 26, 1969)
P. Tournois, assignor to CSF-Companie Generale de Telegraphie, Paris,
ACOUSTICAL DISPERSIVE DELAY LINE HAVING STRATIFIED WAVEGUIDE OF AT LEAST TWO SOLID MEDIA COUPLING INPUT AND OUTPUT TRANSDUCERS.

3,479,572 (Nov. 18, 1969)
G.E. Pokorny, assignor to Litton Precision Products, Inc., San Carlos, Calif.,
ACOUSTIC SURFACE WAVE DEVICE.

3,488,602 (Jan. 6, 1970)
H. Seidel and D.L. White, assignors to Bell Telephone Laboratories, Inc., N.Y.
ULTRASONIC SURFACE WAVEGUIDES.

3,491,291 (Jan. 20, 1970)
C.W.E. Walker, assignor to Beloit Iron Works, Beloit, Wis.,
RESONANT LOOP SINGLE CONDUCTOR SURFACE WAVE DEVICE.

3,500,461 (March 10, 1970)
M. Epstein, D. Fryberger, R.J. Serafin and A.P.van den Heuvel, assignors to IIT Research Institute, Chicago, Ill.,
KINEMAGNETIC PIEZOELECTRIC AND PIEZORESISTIVE TAPPING TECHNIQUES FOR NONMAGNETIC DELAY LINES.

3,504,307 (March 31, 1970)
I.L. Gelles, assignor to Kennecott Copper Corporation, New York,
THIN SAMPLE ULTRASONIC DELAY LINE.

3,505,542 (Apr. 7, 1970)
B.R. McAvoy, assignor to Westinghouse Electric Corporation, Pittsburgh, Pa.,
SURFACE WAVE PIEZOELECTRIC RESONATOR.

1,187,483 (Apr. 8, 1970) British Patent
I.L. Gelles, assignor to Kennecott Copper Corporation, New York,
IMPROVEMENTS IN OR RELATING TO DELAY METHODS AND DELAY LINES FOR THE TRANSMISSION OF ULTRASONIC SIGNALS.
(The equivalent of U.S. Pat. 3,504,307)

3,509,452 (Apr. 28, 1970)
C.W.E. Walker, assignor to Beloit Corporation, Beloit, Wis.,
MICROWAVE HYGROMETER HAVING A HELICAL SURFACE WAVE TRANSMISSION LINE.

3,509,463 (Apr. 28, 1970)
S.N. Watkins, E.D. Woodward and T. Jofuku, assignors to Sylvania Electric Products, Inc., Delaware,
SURFACE WAVE TRANSMISSION SYSTEM.

3,515,911 (June 2, 1970)
G.W. Byram and H.J. Whitehouse, assignors to the United States of America, as represented by the Secretary of the Navy,
SURFACE WAVE TRANSDUCERS.

3,516,027 (June 2, 1970)
J.H. Wasilik, assignor to the United States of America as represented by the Secretary of the Army,
VARIABLE SURFACE-WAVE DELAY LINE.

1,197,860 (July 8, 1970) British Patent
J.S. Heeks, assignor to Standard Telephones and Cables Ltd., London, England,
ACOUSTIC SURFACE WAVE DELAY LINE.

3,523,200 (Aug. 4, 1970)
W.E. Newell, assignor to Westinghouse Electric Corp., Pittsburgh, Pa.,
SURFACE WAVE PIEZOELECTRIC RESONATOR.

3,534,300 (Oct. 13, 1970)
H. Jouffroy and J. Desbois, assignors to Thomson CSF, Paris, France,
DEVICE FOR EXCITING SURFACE WAVES.
(Equivalent to British Patent 1,151,735)

3,548,306 (Dec. 15, 1970)
H.J. Whitehouse, assignor to the United States of America, as represented by the Secretary of the Navy,
SURFACE WAVE SPECTRUM ANALYZER AND INTERFEROMETER.

3,550,045 (Dec. 22, 1970)
R. Adler, assignor to Zenith Radio Corporation, Chicago, Ill.,
ACOUSTIC SURFACE WAVE FILTER DEVICES.

3,551,690 (Dec. 29, 1970)
A. Korpel, assignor to Zenith Radio Corporation, Chicago, Ill.,
WAVE GENERATOR.

3,551,837 (Dec. 29, 1970)
J.M. Speiser and J.H. Whitehouse, assignors to the United States of America as represented by the Secretary of the Navy,
SURFACE WAVE TRANSDUCERS WITH SIDE LOBE SUPPRESSION.

3,559,115 (Jan. 26, 1971)
A.J. DeVries, assignor to Zenith Radio Corporation, Chicago, Ill.,
SURFACE WAVE FILTER REFLECTION CANCELLATION.

3,560,724 (Feb. 2, 1971)
W.J. Condell, Jr, assignor to the United States of America as represented by the Secretary of the Navy,
OPTICAL CORRELATORS.

3,568,102 (Mar. 2, 1971)
C.C. Tseng, assignor to Litton Precision Products, Inc., San Carlos, Calif.,
SPLIT SURFACE WAVE ACOUSTIC DELAY LINE.

3,573,673 (Apr. 6, 1971)
A.J. DeVries and T.J. Wojcik, assignors to Zenith Radio Corporation, Chicago,
ACOUSTIC SURFACE WAVE FILTERS.

3,581,248 (May 25, 1971)
A.J. DeVries and T.J. Wojcik, assignors to Zenith Radio Corporation, Chicago,
ACOUSTIC FILTERS.

3,582,540 (June 1, 1971)
R. Adler, A.J. DeVries and F. Dias, assignors to Zenith Radio Coproration,
SIGNAL TRANSLATING APPARATUS USING SURFACE WAVE ACOUSTIC DEVICE.

3,582,837 (June 1, 1971)
A.J. DeVries, assignor to Zenith Radio Corporation, Chicago, Ill.,
SIGNAL FILTER UTILIZING FREQUENCY-DEPENDENT VARIATION OF INPUT IMPEDANCE OF ONE-PORT TRANSDUCER.

3,582,838 (June 1, 1971)
A.J. DeVries, assignor to Zenith Radio Corporation, Chicago, Ill.,
SURFACE WAVE DEVICES.

3,582,840 (June 1, 1971)
A.J. DeVries, assignor to Zenith Radio Corporation, Chicago, Ill.,
ACOUSTIC WAVE FILTER.

1,234,016 (June 3, 1971) British Patent
Thomson-CSF, Paris, France (Inventor's name not given),
ENERGY WEIGHTED DISPERSIVE ACOUSTIC DELAY LINE OF THE SURFACE WAVE TYPE.

3,587,005 (June 22, 1971)
I.E. Fair, assignor to Bell Telephone Laboratories, Inc., Murray Hill, N.J.,
TRANSDUCER ARRAY FOR ELASTIC WAVE TRANSMISSION.

PATENTS RELATED TO ACOUSTO-OPTIC DEVICES, PUBLISHED BETWEEN 1960 - 1971:

2,943,315 (June 28, 1960)
A.H. Rosenthal, assignor to Fairchild Camera and Instrument Corp., Syosset, N.Y.,
LIGHT MODULATOR DISPLAY MEANS.

3,011,386 (Dec. 5, 1961)
A.H. Rosenthal, assignor to Servo Corp. of America, Hicksville, N.Y.,
SPECTROMETER.

3,012,467 (Dec. 12, 1961)
A.H. Rosenthal, assignor to Servo Corp. of America, Hicksville, N.Y.,
SPECTRUM ANALYZER.

3,035,491 (May 22, 1962)
A.H. Rosenthal and S. Tinto, assignors to Fairchild Camera and Instrument Corp.,
ULTRASONIC LIGHT MODULATOR CELL ASSEMBLY.

3,055,258 (Sept. 25, 1962)
H. Hurvitz, Washington, D.C.,
BRAGG DIFFRACTION ULTRASONIC DEVICES.

3,065,378 (Nov. 20, 1962)
V.Z.Zaks, assignor to The American Society for Technion and Israel Institute of Technology, Inc., New York,
VISUAL DISPLAY AND READOUT SYSTEMS.

3,088,113 (Apr. 30, 1963)
A.H. Rosenthal, assignor to Fairchild Camera and Instrument Corp., Delaware,
CORRELATION SYSTEM FOR RADAR AND THE LIKE.

3,111,666 (Nov. 19, 1963)
R.M. Wilmotte, Princeton, N.J.,
METHOD AND APPARATUS FOR OPTICALLY PROCESSING INFORMATION.

3,121,169 (Feb. 11, 1964)
R.C. Benton, assignor to Centre Circuits, Inc., State College, Pa.,
OPTICAL COMMUNICATION SYSTEMS USING A CIRCULAR ELECTROMECHANICAL MODULATOR.

3,136,893 (June 9, 1964)
W. Liben and L.A. Twigg, Jr., Ellicott City, Md.,
SIGNAL TRANSMISSION SYSTEMS USING SONIC LINES.

3,174,044 (March 16, 1965)
P.K. Tien, assignor to Bell Telephone Laboratories, New York,
LIGHT FREQUENCY MODULATOR.

3,189,746 (June 15, 1965)
L. Slobodin and A. Reich, assignors to Lockheed Aircraft Corp., Burbank, Calif.,
EXPANSION AND COMPRESSION OF ELECTRONIC PULSES BY OPTICAL CORRELATION.

3,205,495 (Sept. 7, 1965)
R.M. Wilmotte, Princeton, N.J.,
OPTICAL PROCESSING OF INFORMATION.

3,220,003 (Nov. 23, 1965)
H. Montague III, J.S. Gerig and G.E. Hoffmann, assignors to Scope Inc.,
DETECTOR FOR VARYING CARRIER FREQUENCY SIGNALS.

3,247,386 (Apr. 19, 1966)
R.C. Vickery, assignor to J.V. Fisher, Valencia, Pa.,
MODULATION OF LASERS BY ULTRASONIC VARIATION OF ABSORPTION BANDS.

3,256,443 (June 14, 1966)
A.R. Moore, assignor to Radio Corporation of America, Delaware,
LASER MULTIPLEX COMMUNICATION SYSTEM.

3,259,014 (July 5, 1966)
L.F. Johnson and D. Kahng, assignors to Bell Telephone Laboratories, Inc., N.Y.
LIGHT MODULATOR USING A VARIABLE SPACING DIFFRACTION GRATING.

3,280,318 (Oct. 18, 1966)
J.S. Gerig, H.C. Galss, G.E. Hoffmann and W.L. Holford, assignors to Scope, Inc.,
Falls Church, Va.,
CORRELATOR.

3,297,876 (Jan. 10, 1967)
A.J. De Maria, assignor to United Aircraft Corporation, East Hartford, Conn.,
AMPLITUDE MODULATION FOR LASERS.

3,306,977 (Feb. 28, 1967)
H.P. Brueggemann, assignor to The Marquardt Corporation, Van Nuys, Calif.,
LINE SCAN DEVICE UTILIZING CAVITATION PHENOMENA PRODUCED IN AN ULTRA-
SONIC CELL.

3,324,478 (June 6, 1967)
J.H. Jacobs, assignor to Bell & Howell Company, Chicago, Ill.,
OSCILLOGRAPHIC RECORDING WITH COHERENT LIGHT.

3,330,956 (July 11, 1967)
G. Wade, assignor to Raytheon Company, Lexington, Mass.,
OPTICAL BEAM MODULATOR USING ACOUSTIC ENERGY.

3,349,344 (Oct. 24, 1967)
A.G. Chynoweth and D.E. McCumber, assignors to Bell Telephone Laboratories, Inc.,
SINGLE VALLEY BULK EFFECT TRAVELING WAVE HIGH FREQUENCY OSCILLATOR, AMPLIFIER
AND MODULATOR.

3,365,579 (Jan. 23, 1968)
M. Emshwiller, assignor to Bell Telephone Laboratories, Inc., New York,
OPTICAL WAVE CORRELATOR WITH ACOUSTICAL MODULATION.

3,365,581 (Jan. 23, 1968)
B. Tell and J.M. Worlock, assignors to Bell Telephone Laboratories, Inc.,
OPTICAL MODULATORS.

3,372,973 (March 12, 1968)
D.E. Flinchbaugh, assignor to United Aircraft Corporation, East Hartford, Conn.,
TWO-DIMENSIONAL BEAM SCANNING DEVICE.

3,373,380 (March 12, 1968)
R. Adler, assignor to Zenith Radio Corporation, Chicago, Ill.,
LASER BEAM SOUNDWAVE TECHNIQUES USING CURVED ACOUSTIC WAVES.

3,383,627 (May 14, 1968)
P.C.J. Desmares, assignor to Zenith Radio Corporation, Chicago, Ill.,
WAVE SELECTOR WITH TAPPED ACOUSTIC TRANSDUCER.

3,383,631 (May 14, 1968)
A. Korpel, assignor to Zenith Radio Corporation, Chicago, Ill.,
ACOUSTIC IMPEDANCE MATCHING.

3,389,348 (June 18, 1968)
A.J. De Maria, assignor to United Aircraft Corporation, East HArtford, Conn.,
GATING SYSTEM FOR HIGH POWER LASER CASCADE.

3,392,368 (July 9, 1968)
R.C. Brewer and K.E. Rieckhoff, assignors to International Business Machines corporation, Armonk, N.Y.
ULTRASONIC AND HYPERSONIC SOUND GENERATOR.

3,397,605 (Aug. 20, 1968)
H.P. Brueggemann, assignor to The Marquardt Corporation, Van Nuys, Calif.,
FREQUENCY MODULATED RADIANT ENERGY SCANNER EMPLOYING CAVITATION-INDUCED DIFFRACTION.

3,397,936 (Aug. 20, 1968)
J.G. Frayne and H.P. Brueggemann, assignors to The Marquardt Corporation, Van Nuys, California,
STANDING WAVE ULTRASONIC LIGHT CELL MODULATOR.

3,399,013 (Aug. 27, 1968)
H.G. Aas and R.K. Erf, assignors to United Aircraft Corp., East Hartford, Conn.,
ULTRASONIC SCANNING CELL.

3,412,269 (Nov. 19, 1968)
E.C. Crittenden, Jr., assignor to TRW, Inc., Redondo Beach, Calif.,
HYPERSONIC TRANSDUCER.

3,413,476 (Nov. 26, 1968)
E.I. Gordon, assignor to Bell Telephone Laboratories, Inc., New York,
LIGHT BEAM CONTROLLING SYSTEM.

3,419,322 (Dec. 31, 1968)
R. Adler, assignor to Zenith Radio Corporation, Chicago, Ill.,
ULTRASONIC TRANSDUCER MATCHING FOR BRAGG REFLECTION SCANNING.

3,421,003 (Jan. 7, 1969)
G.J. Pratt, assignor to Corning Glass Works, Corning, N.Y.
APPARATUS AND METHOD FOR OPTICAL SIGNAL PROCESSING.

3,424,906 (Jan. 28, 1969)
A. Korpel, assignor to Zenith Radio Corporation, Chicago, Ill.,
LIGHT-SOUND INTERACTION SYSTEM WITH ACOUSTIC BEAM STEERING.

3,431,504 (March 4, 1969)
R. Adler, assignor to Zenith Radio Corporation, Chicago, Ill.,
ACOUSTICAL LIGHT-TRANSLATING APPARATUS.

3,432,221 (March 11, 1969)
T.J. Harris and E. Max, assignors to International Business Machines Corp.,
STRESSED LASER SCANNING DEVICE USING LIGHT POLARIZERS.

3,432,222 (March 11, 1969)
H. Fleisher, K.M. Kosanke, W.W. Kulcke and E. Max, assignors to International Business Machines Corporation, New York,
OPTICAL SCANNING DEVICE.

3,432,647 (March 11, 1969)
R.M. Wilmotte, Washington, D.C.,
ELECTRO-OPTICAL CORRELATOR.

3,435,228 (March 25, 1969)
E.I. Gordon, assignor to Bell Telephone Laboratories, Inc., New York,
LIGHT BEAM CONTROLLING SYSTEM.

3,435,372 (March 25, 1969)
H.G. Aas and G.E. Danielson, assignors to United Aircraft Corp., East Hartford,
ULTRASONIC MODULATOR HAVING A CYLINDRICAL TRANSDUCER.

3,441,724 (Apr. 29, 1969)
D.R. Taylor, Jr., assignor to Philco-Ford Corp., Philadelphia, Pa.,
OPTICAL CORRELATOR.

1,150,625 (Apr. 30, 1969) British Patent
S.L. Boersma, Delft, The Netherlands,
ELECTRONIC IMAGE TRANSMISSION SYSTEM.

3,446,967 (May 27, 1969)
J.S. Palfreeman, assignor to U.S. Philips Corporation, New York,
APPARATUS FOR REVERSING THE SEQUENCE OF PULSES IN A PULSE GROUP.

3,453,595 (July 1, 1969)
E.S. Barrekette, C.V. Freiman and D.C. Gazis and J.E. Lovell, assignors to International Business Machines Corporation, New York,
OPTIC TO ACOUSTIC CONVERTER FOR PATTERN RECOGNITION.

3,453,626 (July 1, 1969)
D.L. Wilson and G.R. Knight, assignors to Sylvania Electric Products, Inc.,
ACOUSTIC GONIOMETER.

3,457,425 (July 22, 1969)
K. Preston, Jr., assignor to The Perkin-Elmer Corporation, Norwalk, Conn.,
OPTICAL CORRELATOR MODULATING LIGHT TWICE IN ULTRASONIC LIGHT MODULATOR.

3,462,212 (Aug. 19, 1969)
R.T. Denton, assignor to Bell Telephone Laboratories, Inc., New York,
POLYCHROMATIC MEAN DEFLECTION.

3,462,603 (Aug. 19, 1969)
E.I. Gordon, assignor to Bell Telephone Laboratories, Inc., New York,
ACOUSTIC LIGHT MODULATOR AND VARIABLE DELAY DEVICE.

3,462,714 (Aug. 19, 1969)
R. Adler, assignor to Zenith Radio Corporation, Chicago, Ill.,
ACOUSTIC SIGNAL TRANSLATING DEVICE HAVING A PROPAGATING MEDIUM COMPOSED OF A PLURALITY OF EFFECTIVELY DISTINCT SIGNAL TRANSLATING PATHS OF MUTUALLY DIFFERENT EFFECTIVE LENGTHS.

3,463,573 (Aug. 26, 1969)
M.J. Brienza, assignor to United Aircraft Corporation, East Hartford, Conn.,
CONTINUOUSLY VARIABLE LASER ACOUSTIC DELAY LINE.

3, 464,027 (Aug. 26, 1969)
A.J. De Maria, assignor to United Aircraft Corporation, East Hartford, Conn.,
LASER MODULATION BY FOCUSED ACOUSTIC ENERGY.

1,166,892 (Oct. 15, 1969) British Patent
The Perkin-Elmer Corporation, Norwalk, Conn.,
OPTICAL CORRELATOR.
(Equivalent of U.S. Pat. 3,457,425)

3,479,494 (Nov. 18, 1969)
R.M. Wilmotte, Washington, D.C.,
INFORMATION PROCESSING AND DISPLAY WITH OPTICAL CORRELATION.

1,171,411 (Nov. 19, 1969) British Patent
N.S. Shiren and A.W. Smith, assignors to International Business Machines Corporation, New York,
OPTICAL MODULATOR.

3,482,101 (Dec. 2, 1969)
F.H. Slaymaker, assignor to General Dynamics Corporation, Delaware,
ELECTRO-OPTICAL SIGNAL PROCESSING SYSTEMS.

3,483,386 (dec. 9, 1969)
J.L. Jernigan, assignor to the United States of America as represented by the Secretary of the Navy,
BRAGG ANGLE MATCHED FILTER FOR PRODUCING AND RECOGNIZING CODED PULSES.

3,483,387 (Dec. 9, 1969)
L. Davis, Jr., assignor to the Raytheon Company, Lexington, Mass.,
ULTRASONIC OPTICAL MODULATOR FOR TIME COMPRESSION OF CHIRP SIGNALS.

3,483,942 (Dec. 16, 1969)
M.H. Crowell and D. Maydan, assignors to Bell Telephone Laboratories, Inc., N.Y.,
ACOUSTIC DEVICES.

3,485,552 (Dec. 23, 1969)
R. Adler, assignor to Zenith Radio Corporation, Chicago, Ill.,
PHASED ARRAY-TYPE BEAM SCANNER WITH DISPERSION COMPENSATION.

3,485,559 (Dec. 23, 1969)
A.J. De Maria, assignor to United Aircraft Corporation, East Hartford, Conn.,
ANGLE MEASURING APPARATUS UTILIZING LASERS.

3,488,437 (Jan. 6, 1970)
A. Korpel, assignor to Zenith Radio Corporation, Chicago, Ill.,
VIDEO DISPLAY SYSTEMS.

3,488,438 (Jan. 6, 1970)
A. Korpel, assignor to Zenith Radio Corporation, Chicago, Ill.,
DISPLAY SYSTEM UTILIZING BRAGG DIFFRACTION.

3,488,503 (Jan. 6, 1970)
A. Reich, R. Lipnick and G.A. Schoen, assignors to Lockheed Aircraft Corporation, Burbank, Calif.,
ACOUSTIC LIGHT SCANNER AND DISPLAY SYSTEM.

3,492,495 (Jan. 27, 1970)
A.J. De Maria and M.J. Brienza, assignors to United Aircraft Corporation, East Hartford, Conn.,
FREQUENCY TRANSLATED LASER ACOUSTIC DELAY LINE.

3,492,596 (Jan. 27, 1970)
G.C. Vorie, assignor to Cornell Aeronautical Laboratory, Inc., Buffalo, N.Y.,
LASER SCANNER.

3,493,759 (Feb. 3, 1970)
R. Adler, assignor to Zenith Radio Corporation, Chicago, Ill.,
ACOUSTIC BEAM STEERING WITH ECHELON TRANSDUCER ARRAY.

1,180,878 (Feb. 11, 1970) British Patent
United Aircraft Corporation, East Hartford, Connecticut,
LASER-ACOUSTIC SIGNAL PROCESSOR.

3,495,894 (Feb. 17, 1970)
P.V. Lenzo, K. Nassau and E.G. Spencer, assignors to Bell Telephone Laboratories, Inc., New York,
OPTICAL BEAM DEFLECTION UTILIZING $LiNbO_3$ OR $LiTaO_3$.

3,499,705 (March 10, 1970)
A.C. Baynham and H.D. Rees, assignors to Minister of Technology in Her Britannic Majesty's Government of the United Kingdom of Great Britain and Northern Ireland, London, England,
TRAVELLING SPOT DISPLAY DEVICE.

3,502,391 (March 24, 1970)
I.P. Kaminow, assignor to Bell Telephone Laboratories, Inc., Murray Hill, N.J.,
OPTICAL BEAM DEFLECTOR USING DIVERGING BEAMS.

3,502,879 (March 24, 1970)
L.M. Vallese, assignor to International Telephone and Telegraph Corp., Delaware,
LIGHT SCANNING DEVICE.

3,506,334 (Apr. 14, 1970)
A. Korpel, assignor to Zenith Radio Corporation, Chicago, Ill.,
PHASED ARRAY-TYPE BEAM SCANNING.

3,506,928 (Apr. 14, 1970)
A. Korpel, assignor to Zenith Radio Corporation, Chicago, Ill.,
RADIATION SCANNING APPARATUS.

3,507,553 (Apr. 21, 1970)
L.K. Anderson and S.S. Devlin, assignors to Bell Telephone Laboratories, Inc.,
ULTRASONIC DIGITAL LIGHT DEFLECTOR.

3,509,453 (Apr. 28, 1970)
R.M. Wilmotte, Princeton, N.J.,
LIGHT MODULATION SYSTEM FOR ANALYSIS OF INFORMATION.

3,509,489 (Apr. 28, 1970)
G. Burns and A. W. Smith, assignors to International Business Machines Corp.,
LASER DEFLECTION.

3,509,565 (Apr. 28, 1970)
R.M. Wilmotte, Princeton, N.J.,
METHOD AND APPARATUS FOR OPTICALLY PROCESSING INFORMATION.

3,512, 871 (May 19, 1970)
H.J. Zweig, assignor to International Business Machines Corporation, New York,
LIGHT BEAM DEFLECTION USING FOURIER OPTICS.

1,192,491 (May 20, 1970) British Patent
Mullard Ltd., London, England,
APPARATUS FOR TREATING A LIGHT BEAM.

3,514,534 (May 26, 1970)
A. Korpel, assignor to Zenith Radio Corporation, Chicago, Ill.,
WIDE-APERTURE LASER BEAM MODULATING AND SCANNING SYSTEM.

3,515,458 (June 2, 1970)
A. Korpel and R.L. Whitman, assignors to Zenith Radio Corporation, Chicago, Ill.,
RADIATION SCANNING SYSTEM.

3,516,729 (June 23, 1970)
R. Adler, assignor to Zenith Radio Corporation, Chicago, Ill.,
CYLINDRICAL LENS COMPENSATION OF WIDE-APERTURE BRAGG DIFFRACTION SCANNING CELL.

3,517,332 (June 23, 1970)
A.J. De Maria, assignor to United Aircraft Corporation, East Hartford, Conn.,
FREQUENCY MODULATION FOR LASERS.

3,524,011 (Aug. 11, 1970)
A. Korpel, assignor to Zenith Radio Corporation, Chicago, Ill.,
LASER COLOR DISPLAY UTILIZING ACOUSTICAL LIGHT MODULATORS.

3,529,263 (Sept. 15, 1970)
W.W. Simmons, assignor to TRW, Inc., Redondo Beach, Calif.,
MEANS FOR EXTRACTING LIGHT FROM MODE-LOCKED LASERS.

3,529,886 (Sept. 22, 1970)
R.W. Dixon, and D.A. Pinnow, assignors to Bell Telephone Laboratories, Inc.,
IODIC ACID ACOUSTO-OPTIC DEVICES.

3,530,400 (Sept. 22, 1970)
G.W. Pratt, Jr., and J.E. Ripper, assignors to Massachusetts Institute of Technology, Cambridge, Mass.,
ACOUSTICALLY MODULATED LASER.

3,531,184 (Sept. 29, 1970)
R. Adler, assignor to Zenith Radio Corporation, Chicago, Ill.,
MONOCHROMATIC LIGHT BEAM DEFLECTION APPARATUS HAVING TWO TRAINS OF FREQUENCY SCANNED ACOUSTIC WAVES FOR EFFECTING BRAGG DIFFRACTION.

3,534,166 (Oct. 13, 1970)
A. Korpel, assignor to Zenith Radio Corporation, Chicago, Ill.,
TELEVISION PICTURE RECORDING AND REPLAY SYSTEM.

3,536,376 (Oct. 27, 1970)
H.B. Henning, assignor to Raytheon Company, Lexington, Mass.,
OPTICAL MODULATION APPARATUS HAVING A PHASE RETARDING SPATIAL FILTER.

3,539,244 (Nov. 10, 1970)
G.E. Marks, assignor to Sperry Rand Corporation, New York,
ELECTROMAGNETIC RADIATION DEFLECTION APPARATUS EMPLOYING ELECTRO-OPTIC OR ELECTRO-ACOUSTICAL DEVICES.

3,539,245 (Nov. 10, 1970)
M.J. Brienza, assignor to United Aircraft Corporation, East Hartford, Conn.,
LASER-ACOUSTIC SIGNAL PROCESSOR.

3,544,795 (Dec. 1, 1970)
A. Korpel, assignor to Zenith Radio Corporation, Chicago, Ill.,
ELECTRO-OPTICAL SIGNAL TRANSFER APPARATUS.

3,544,805 (Dec. 1, 1970)
A.J. De Maria, assignor to United Aircraft Corporation, East Hartford, Conn.,
LASER DELAY LINE USING BIASING SIGNAL.

3,544,806 (Dec. 1, 1970)
A.J. De Maria and M.j. Brienza, assignors to United Aircraft Corp., East Hartford,
CONTINUOUSLY VARIABLE LASER-ACOUSTIC DELAY LINE.

1,216,955 (Dec. 23, 1970) British Patent
M.J. Debtino and E.A. Reitz, assignors to North American Rockwell Corporation, El Segundo, California,
NON-SCANNING OPTICAL PROCESSOR FOR DISPLAY OF VIDEO SIGNALS.

3,562,414 (Feb. 9, 1971)
A.S. Blum, assignor to Zenith Radio Corporation, Chicago, Ill.,
SOLID-STATE IMAGE DISPLAY DEVICE WITH ACOUSTIC SCANNING OF STRAIN-RESPONSIVE SEMICONDUCTOR.

3,566,303 (Feb. 23, 1971)
A.J. De Maria, assignor to United Aircraft Corporation, East Hartford, Conn.,
ULTRASONIC CONTROL SYSTEM FOR LASERS.

3,571,507 (March 16, 1971)
A. Korpel, Assignor to Zenith Radio Corporation, Chicago, Ill.,
OPTICAL APPARATUS FOR DEVELOPING DISPLAY INFORMATION SIGNALS OF FREQUENCY MULTIPLEX CHARACTER.

3,572,899 (March 30, 1971)
deLamar T. Bell, Jr., assignor to Texas Instruments, Inc., Dallas, Texas,
ACOUSTO-OPTICAL MODULATOR.

3,575,550 (Apr. 20, 1971)
A. Korpel, assignor to Zenith Radio Corporation, Chicago, Ill.,
OPTICAL APPARATUS FOR DEVELOPING A FREQUENCY-DOMAIN SIGNAL.

3,582,635 (June 1, 1971)
L. Slobodin, assignor to Lockheed Aircraft Corporation, Burbank, California,
OPTICAL CORRELATION OF ELECTRIC PULSES.

3,585,392 (June 15, 1971)
A. Korpel, assignor to Zenith Radio Corporation, Chicago, Ill.,
PHASE MODULATED LIGHT DETECTION.

3,585,848 (June 22, 1971)
A. Korpel, assignor to Zenith Radio Corporation, Chicago, Ill.,
METHOD AND APPARATUS FOR RECORDING ACOUSTIC IMAGES AND HOLOGRAMS.

****

PATENT INFORMATION

Regarding U.S. Patents: Copies of U.S. Patents may be obtained from:

The Commissioner of Patents
Washington, D.C. 20231

$.50 per copy paid in advance by Post Office Money Order or Certified Check (made payable to The Commissioner of Patents).

Regarding British Patents: To obtain complete patent specifications address request to: Comptroller-General, The Patent Office (Sale Branch), St. Mary Cray, Orpington, Kent BR5 3RD, England. 25p per copy payable in International Money Order or bank draft. (Payable to The Comptroller General, The Patent Office).

****